BIOTECHNOLOGY TOOLS FOR CROP BREEDING

BIOTECHNOLOGY TOOLS FOR CROP BREEDING

By

Prof. Leena Wong

2016

SBS Publishers & Distributors Pvt. Ltd.
New Delhi

ISBN 13 : 9789380090696

First Published in 2016

Published by:

SBS PUBLISHERS & DISTRIBUTORS PVT. LTD.

2/9, Ground Floor, Ansari Road, Darya Ganj,

New Delhi - 110002,

INDIA

Tel: 0091.11.23289119 / 41563911

Email: mail@sbspublishers.com

www.sbspublishers.com

Preface

Plant breeding is the art and science of changing the traits of plants in order to produce desired characteristics. Plant breeding can be accomplished through many different techniques ranging from simply selecting plants with desirable characteristics for propagation. Plant Breeding is the scientific improvement of plants to provide an economically and environmentally sustainable food, feed and fiber supply. Our plant breeding program use both phenotypic and molecular assisted selection techniques and develops new methodologies to phenotype and improve plants to improve quality, output potential of grain, minimize yield losses and reduce crop production costs. Our plant breeding program alters the genetic composition of varieties to improve crop characteristics. Relying on breeding and molecular techniques for specific target goals to improve quality, output potential of grain, minimizes yield losses and reduce crop production costs. Classical plant breeding uses deliberate interbreeding (crossing) of closely or distantly related individuals to produce new crop varieties or lines with desirable properties. Plants are crossbred to introduce traits/genes from one variety or line into a new genetic background. Modern plant breeding may use techniques of molecular biology to select, or in the case of genetic modification, to insert, desirable traits into plants. Application of biotechnology or molecular biology is also known as molecular breeding. Biotechnology in its broadest sense refers to the use of living organisms or their components to provide useful products. This definition can include activities as diverse as making wine, beer, or bread; composting organic materials; releasing parasitic wasps to control insect pests; breeding plants or animals; and producing crops and livestock. In fact, agriculture itself can be considered to be the original biological technology. With increasing knowledge of genetics, plant breeders in this century have accelerated the improvement of crops for enhanced yield and quality. The molecular breeding program utilizes DNA marker technology to aid the efficient and effective incorporation of valuable traits into improved rice cultivars. Many traits important in the development of cultivars can be difficult to measure or are environmentally sensitive, which makes it difficult to make progress in breeding programs. DNA marker technology aids in overcoming these limitations by using a small amount of seed or leaf tissue from a plant to perform an easily interpreted laboratory test for indicating the likelihood of having the trait of interest. Because it is difficult to combine numerous important traits simultaneously into cultivars, marker-assisted breeding gives geneticists a versatile set of tools that can augment, as well as verify, traditional selection techniques.

Editor

Contents

Chapter 1

DEVELOPMENT OF DART MARKER PLATFORMS AND GENETIC DIVERSITY ASSESSMENT OF THE U.S. COLLECTION OF THE NEW OILSEED CROP LESQUERELLA AND RELATED SPECIES

[1]Von Mark V. Cruz, [2]Andrzej Kilian, [1]David A. Dierig

[1]USDA-ARS National Center for Genetic Resources Preservation, Fort Collins, Colorado, United States of America, Department of Bioagricultural Sciences and Pest Mgt., Colorado State University, Fort Collins, Colorado, United States of America

[2]Diversity Arrays Technology Pty. Ltd., Yarralumla, ACT, Australia

ABSTRACT

The advantages of using molecular markers in modern genebanks are well documented. They are commonly used to understand the distribution of genetic diversity in populations and among species which is crucial for efficient management and effective utilization of germplasm collections. We describe the development of two types of DArT molecular marker platforms for the new oilseed crop lesquerella (*Physaria* spp.), a member of the Brassicaceae family, to characterize a collection in the National Plant Germplasm System (NPGS) with relatively little known in regards to the genetic diversity and traits. The two types of platforms were developed using a subset of the germplasm conserved *ex situ* consisting of 87 *Physaria* and 2 *Paysonia*accessions. The microarray DArT revealed a total of

2,833 polymorphic markers with an average genotype call rate of 98.4% and a scoring reproducibility of 99.7%. On the other hand, the DArTseq platform developed for SNP and DArT markers from short sequence reads showed a total of 27,748 high quality markers. Cluster analysis and principal coordinate analysis indicated that the different accessions were successfully classified by both systems based on species, by geographical source, and breeding status. In the germplasm set analyzed, which represented more than 80% of the *P. fendleri* collection, we observed that a substantial amount of variation exists in the species collection. These markers will be valuable in germplasm management studies and lesquerella breeding, and augment the microsatellite markers previously developed on the taxa.

INTRODUCTION

Lesquerella (*Physaria* sp.), a member of the Brassicaceae, is a North American genus which has more than 90 member species thriving mostly in dry and arid habitats usually mixed with sparse vegetation [1], [2]. Commonly known as bladderpod, it has been identified by the U.S. Department of Agriculture and the U.S. Department of Energy as very promising species for the production of lubricants, engine oils, waxes, coatings and a source of natural estolides with valuable application in the automobile and biofuel industries [3], [4], [5]. There has been an increasing trend in research activities in the crop since its oil was first characterized for novel properties [6]. The fatty acid profile of lesquerella oil has been found to vary. Species found east of the Mississippi have mostly densipolic acid, those in the western U.S. mostly lesquerolic acid, and a species from Oklahoma predominantly with auricolic acid [7]. Lesquerella could be grown successfully as a winter annual in the southwest U.S. producing an average seed yield of 1.7 tons/ha [8]. At present, there are more than eight advanced breeding lines of *P. fendleri*ready for commercialization and a corresponding substantial germplasm collection has been assembled in the U.S. National Plant Germplasm System [9].

Molecular characterization of germplasm collections supplements phenotypic assessment of diversity and is important in the effective management of genetic resources. Molecular markers have been very useful in efforts to accurately identify gaps and redundancy within and among individual germplasm collections and have helped resolve important genebank management issues [10], [11], [12]. Examples of markers used in specific Brassicaceae collections include microsatellites for examining diversity in *B. napus* [13], [14], [15], lesquerella [16] and wild relatives such as *Capsella, Crambe* and *Sinapis* [17]; AFLPs in *B. oleracea* [18] and *Lepidium*[19]; and RAPD markers in *Raphanus* [20].

Compared to the previously mentioned molecular marker systems, Diversity Array Technology (DArT) markers are relatively new and were developed only in the early 2000. DArT markers overcame the difficulties in correlating bands on gels with allelic variants by utilizing hybridization-based methods and solid state surfaces [21]. They have been increasingly utilized as evidenced by the number of publications reporting successful implementation [22]. Among its advantages over other marker systems include high throughput capability allowing rapid germplasm characterization in a single experiment, independence of sequence data, and ability to detect single base changes and indels [23]. To date, DArT markers have been successfully applied in genetic diversity analysis, linkage mapping and in finding out population structure of collections in various crop species [24]. Their application in minor crops are likewise increasing due to their potential to accelerate gene discovery and initiate molecular breeding because of their whole genome coverage without relying on prior sequence data information [25], [26].

In lesquerella, allozymes have previously been used [27] and microsatellite markers have been developed [16]. But the small number of markers that are available presents a limitation in linkage mapping and in the study of genetic resources collections. At present, only fifteen microsatellites have demonstrated utility across different *Physaria* species. We present in this paper the development of two platforms of high density DArT markers for lesquerella and the results obtained after testing them to analyze the genetic diversity of *Physaria* and *Paysonia*national germplasm collection. This new molecular marker system for the new crop species will serve as an additional resource to augment the existing systems to assist crop improvement efforts, germplasm management activities and genetic studies.

MATERIALS AND METHODS

Tissue Sampling

The lesquerella germplasm set was obtained from collections of the USDA-ARS National Arid Land Plant Genetic Resources Unit, Parlier, CA and the USDA-ARS Arid Land Agricultural Research Center, Maricopa, AZ. The samples used in this study were selected based on each accession's geographic location, sampling within distinct counties in each State when possible. Seven advanced *P. fendleri* breeding lines, WCL-LH1, WCL-LO1, WCL-LO2, WCL-LO4, WCL-LY1, WCL-SL1 and WCL-YS1 were included among the samples. DNA was extracted from fresh leaf tissue obtained from five week old seedlings using Qiagen DNeasy 96 Plant Kits (Qiagen Inc., Valencia, CA). Two DArT platforms were developed for lesquerella as described below. The platform development utilized 86 *Physaria* accessions (see Table 1), representing 11 species and one *Paysonia* accession since the latter is a sister genus of *Physaria* and only recently was there a recognized taxonomic delineation between them [2].

Table 1 Passport information of accessions used in Physaria DArT and DArTseq platform development.

Collection No. and ID	Accession No.	Species	Country	State/ Prov.	County	Latitude	Longitude	Elevation (m)
1809	WS 20822	*P. fendleri*	United States	AZ	Cochise	31.75000	−110.00500	154.15
1817	WS 20823	*P. fendleri*	United States	AZ	Cochise	31.74100	−109.90800	1493.90
1818	WS 20824	*P. fendleri*	United States	AZ	Cochise	31.50000	−110.00000	1432.56
1819	WS 20825	*P. fendleri*	United States	AZ	Santa Cruz	31.72400	−110.52600	1493.90
1826	WS 20856	*P. fendleri*	United States	AZ	Santa Cruz	31.72300	−110.52400	1463.41
1833	WS 20833	*P. gordonii*	United States	NM	Hidalgo	31.93333	−108.95000	1024.13
1834	PI 641918	*P. fendleri*	United States	NM	Hidalgo	31.93333	−108.93333	1292.35
1835	–	*P. fendleri*	United States	NM	Luna	32.21667	−107.41667	1277.11
1840	WS 20858	*P. fendleri*	United States	NM	Eddy	32.61667	−104.40000	1042.42
1841	PI 596417	*P. fendleri*	United States	NM	Eddy	32.31667	−104.25000	1130.81
1842	WS 20860	*P. fendleri*	United States	NM	Eddy	32.53333	−103.98333	1060.70
1843	PI 596418	*P. fendleri*	United States	NM	Eddy	32.75000	−104.10000	1082.04
1844	PI 596419	*P. fendleri*	United States	NM	Lincoln	33.68333	−105.83333	1636.78
1845	PI 596420	*P. fendleri*	United States	NM	Lincoln	33.68333	−105.91667	1572.77
1848	PI 596421	*P. fendleri*	United States	AZ	Graham	33.25000	−110.28333	795.53
1851	WS 20836	*P. fendleri*	United States	AZ	Cochise	31.91667	−109.16667	1493.52
1852	WS 20861	*P. fendleri*	United States	AZ	Cochise	31.65000	−110.01667	1447.80
1874	PI 596423	*P. fendleri*	United States	AZ	Apache	34.56667	−109.50000	1755.65
1889	PI 596425	*P. fendleri*	United States	NM	Socorro	33.88333	−106.40000	1661.16
1904	WS 20865	*P. fendleri*	United States	AZ	Apache	34.35000	−109.36667	1859.28
1906	PI 596426	*P. fendleri*	United States	NM	Socorro	34.41667	−106.65000	1645.92
1909	WS 20868	*P. fendleri*	United States	NM	Guadalupe	33.56667	−105.05000	1697.74
1910	PI 596428	*P. fendleri*	United States	NM	De Baca	34.46667	−104.30000	1149.10
1912	PI 596429	*P. fendleri*	United States	NM	Guadalupe	34.73333	−104.46667	1359.41
1919	PI 596430	*P. fendleri*	United States	NM	San Juan	36.60000	−108.91667	1758.70
1920	PI 596431	*P. fendleri*	United States	NM	San Juan	36.58333	−108.96667	1932.43
1924	WS 20851	*P. recutipes*	United States	AZ	Luna	33.86667	−109.60000	1912.00
1932	PI 596433	*P. fendleri*	United States	AZ	Graham	33.28333	−110.38333	877.82
2212	WS 20863	*P. argyraea*	United States	TX	Uvalde	29.15000	−99.70000	228.60
2226	WS 20828	*P. fendleri*	United States	TX	Webb	27.36667	−98.98333	243.84
2232	PI 643174	*P. lindheimeri*	United States	TX	Nueces	27.06667	−97.06667	107.00
2243	WS 20835	*Paysonia grandiflora*	United States	TX	Wilson	29.10667	−97.93333	91.44
2250	WS 20854	*P. recurvata*	United States	TX	Medina	29.53333	−99.21667	434.34
2255	PI 596435	*P. fendleri*	United States	TX	El Paso	31.83333	−106.01667	1341.12
2256	PI 596436	*P. fendleri*	United States	TX	Hudspeth	31.70000	−105.41667	1240.54
2260	PI 596439	*P. fendleri*	United States	TX	Hudspeth	31.81667	−105.13333	1219.20
2262	PI 596441	*P. fendleri*	United States	TX	Culberson	31.36667	−104.81667	944.88
2269	PI 596447	*P. fendleri*	United States	TX	Pecos	30.71667	−103.18333	975.36
2273	PI 596450	*P. fendleri*	United States	TX	Brewster	30.20000	−103.56667	1554.48
2277	PI 596452	*P. fendleri*	United States	TX	Brewster	29.78333	−103.18333	853.44
2278	PI 596453	*P. fendleri*	United States	TX	Brewster	30.20000	−103.20000	1310.64
2281	PI 596454	*P. fendleri*	United States	TX	Crockett	30.93333	−101.86667	640.08
2282	PI 596455	*P. fendleri*	United States	TX	Pecos	30.96667	−101.96667	563.88
2286	PI 596456	*P. fendleri*	United States	TX	Andrews	32.31667	−102.50000	807.72
2290	PI 596457	*P. fendleri*	United States	TX	Winkler	31.78333	−103.26667	853.44
2291	PI 596458	*P. fendleri*	United States	TX	Reeves	31.11667	−103.71667	944.88
2292	PI 596459	*P. fendleri*	United States	TX	Reeves	31.01667	−103.73333	1005.84
2294	PI 596461	*P. fendleri*	United States	TX	Jeff Davis	30.53333	−104.10000	1722.12
2295	PI 596462	*P. fendleri*	United States	TX	Jeff Davis	30.55000	−103.91667	1676.40
2296	PI 596463	*P. fendleri*	United States	TX	Jeff Davis	30.43333	−103.96667	1554.48
2299	PI 596464	*P. fendleri*	United States	TX	Presidio	30.15000	−104.08333	1402.08
2300	PI 596465	*P. fendleri*	United States	TX	Brewster	30.31667	−103.78333	1493.52
2301	PI 596466	*P. fendleri*	United States	TX	Presidio	30.26667	−103.90000	1432.56
2302	PI 596467	*P. fendleri*	United States	TX	Jeff Davis	30.53333	−104.31667	1493.52
2303	PI 596468	*P. fendleri*	United States	TX	Jeff Davis	30.71667	−104.66667	1280.16
2304	WS 20872	*P. fendleri*	United States	TX	Culberson	30.83333	−104.78333	1188.72
2305	WS 20830	*P. fendleri*	United States	TX	Hudspeth	31.03333	−104.90000	1310.64
2928	WS 20834	*P. gracilis*	United States	OK	Choctaw	34.01667	−95.35000	137.00
3009	WS 20319	*Paysonia auriculata*	United States	OK	Caddo	35.54173	−98.57041	475.00
4005	PI 643177	*P. fendleri*	Mexico	COA	–	28.73861	−100.91250	330.00
4008	–	*P. fendleri*	Mexico	COA	–	28.69528	−100.57672	270.97
4015	–	*P. fendleri*	Mexico	COA	–	25.51048	−101.11323	2007.00
4016	PI 643179	*P. fendleri*	Mexico	ZAC	–	24.37749	−101.38860	1840.00
4024	PI 641924	*P. fendleri*	Mexico	DUR	–	24.59139	−103.92303	1935.00
4042	–	*P. fendleri*	Mexico	COA	–	26.49582	−100.70158	312.72
4043	–	*P. fendleri*	Mexico	COA	–	26.53963	−100.46233	271.27
4044	–	*P. fendleri*	Mexico	ZAC	–	24.24713	−101.43585	1897.68
4045	–	*P. fendleri*	Mexico	ZAC	–	24.23212	−101.44367	1908.05
4046	–	*P. fendleri*	Mexico	ZAC	–	24.26105	−101.44598	1909.27
4047	–	*P. fendleri*	Mexico	NLE	–	25.10865	−100.65158	1919.94
4048	–	*P. fendleri*	Mexico	ZAC	–	24.21158	−101.45683	1935.18
4049	–	*P. fendleri*	Mexico	ZAC	–	24.39118	−101.45952	1933.35
4050	–	*P. fendleri*	Mexico	ZAC	–	24.19929	−101.46171	1930.91
4058	–	*P. fendleri*	United States	NM	McKinley	36.44348	−108.13162	2232.66
4085	–	*P. fendleri*	United States	NM	San Miguel	35.50000	−104.25385	1368.00
4087	–	*P. kathyrn*	United States	ME	–	–	–	–
4092	PARL 812	*P. pallida**	United States	TX	–	–	–	–
4094	PARL 814	*P. pallida**	United States	TX	–	–	–	–
4057b	–	*P. fendleri*	United States	NM	McKinley	–	–	–
DDMC2010-1	PARL 859	*P. fendleri*	United States	NM	Socorro	34.13115	−106.90907	1424.00
DDMC2010-19	PARL 874	*P. fendleri*	United States	TX	Jeff Davis	30.79047	−104.16759	1668.00
DDMC2010-6	PARL 863	*P. gordonii*	United States	TX	Hudspeth	31.74230	−105.10552	1141.00
WCL-LO2	–	*P. fendleri*	United States	–	–	–	–	–
WCL-LO4		*P. fendleri*	United States	–	–	–	–	–
WCL-SL1	PI 673132	*P. fendleri*	United States	–	–	–	–	–
WCL-YS1	PI 610492	*P. fendleri*	United States	–	–	–	–	–
–	PARL 817	*P. thamnophila*	United States	TX	–	–	–	–

*exact geographic coordinates of *P. pallida* are not supplied since they are in the federal and state list of endangered species.

doi:10.1371/journal.pone.0064062.t001

doi:10.1371/journal.pone.0064062.t001

MICROARRAY DART PLATFORM DEVELOPMENT

The microarray DArT was developed by first testing combinations of the rare-cutting restriction enzyme *PstI* with several restriction endonucleases that cut frequently on DNA samples from 8 representative accessions to determine the restriction enzyme combination that provided the best complexity reduction. Final genomic representations were prepared using *PstI/BstNI*combinations. Approximately 50 ng of genomic DNA was digested with *PstI/BstNI* combinations and the resulting fragments ligated to a *PstI* overhang compatible oligonucleotide adapter. A primer annealing to this adapter was used in PCR reaction to amplify complexity-reduced representation of a sample. Amplification products were either used for cloning a in marker development process or labeled with fluorescent dyes and hybridized to DArT array in the genotyping process. Library construction was subsequently performed using 80 *P. fendleri*accessions and 16 accessions of wild related *Brassica* species. The amplified *PstI* restriction fragments from all accessions were cloned into pCR2.1-TOPO vector (Invitrogen, Australia) as described by Jaccoud et al. [21] and four libraries were generated. The white colonies containing genomic fragments inserted into pCR2.1-TOPO vector were picked into individual wells of 384-well microtiter plates filled with ampicillin/kanamycin-supplemented freezing medium. A total of 3,456 clones from *Physaria* were obtained –1,920 from three libraries of *P. fendleri*and 1,536 from the related species. Inserts from these clones were amplified using M13F and M13R primers in 384 plate format, a subset of PCR products were assessed for quality (10% of 25 μl PCR reaction) through gel electrophoresis, and all remaining PCR products dried, washed and dissolved in a spotting buffer. A total of 6,144 clones were printed with spot duplication on SuperChip poly-L-lysine slides (Thermo Scientific, Australia) using a MicroGrid arrayer (Genomics Solutions, UK). The microarrays included 1,920 *Brassica* clones and 768*Arabidopsis* clones in addition to the *Physaria* clones.

Each sample was assayed using methods described above for library construction. Genomic representations were assessed for quality through gel electrophoresis in 1.2% agarose and labeled with fluorescent dyes (Cy3 and Cy5). Labelled targets were then hybridized to printed DArT arrays for 16 hours at 62°C in a water bath. Slides were washed as described by Kilian et al. [28], dried initially by centrifugation at 500 × g for 7 min and later by a desiccator under vacuum for 30 min. The slides were scanned using Tecan LS300 scanner (Tecan Group Ltd, Männedorf, Switzerland) generating three images per array: one image scanned at 488 nm for reference signal measures the amount of DNA within the spot based on hybridization signal of FAM-labeled fragment of a TOPO vector multiple cloning site fragment and two images for "target" signal measurement: one scanned at 543 nM (for Cy3 labeled targets) and one at 633 nM (for Cy5 labeled targets). All DArT techniques applied for this work were recently described in much more detail by Kilian et al. [28].

DARTSEQ PLATFORM DEVELOPMENT

For the sequencing-based DArT genotyping, four complexity reduction methods optimized for several other plant species at DArT P/L were used. *PstI*-RE site specific adapter was tagged with 96 different barcodes enabling encoding a plate of DNA samples to run within a single lane on an Illumina Genome Analyzer IIx (Illumina Inc., San Diego, CA). *PstI* adapter included also a sequencing primer site, so that the tags generated were always reading into the genomic fragments from the *PstI* sites. After the sequencing run, the FASTQ files were quality filtered using the threshold of 90% confidence for at least 50% of the bases and in addition filtered more stringently for barcode sequences. Two lanes of GAIIx were run with all samples providing fully replicated sequencing data. The filtered data were split into their respective target (individual) data using barcode splitting script. Each sample had on average 500,000 counts per replicate. After producing various QC statistics and trimming of the barcode the sequences were aligned against the reference created from the tags identified in the sequence reads generated from all the samples. In addition the short sequence tags were aligned against *Arabidopsis thaliana*'s genome available in Genbank. *Arabidopsis* is a close relative of *Physaria* in thc Brassicaceae [29]. The output files from alignment generated using Bowtie software [30] were processed using an analytical pipeline developed by DArT P/L to produce "DArT score" tables and "SNP" tables.

Genotyping

Both DArT on array platform and DArTseq use a set of quality parameters to select markers which are of use for a specific application. One of these parameters is reproducibility of markers in technical replicates for a subset of samples. In diversity analysis, the reproducibility parameter threshold is set usually at 97% which translates to average reproducibility of the dataset around 99.7%. A total of 87 common accessions were genotyped using both the microarray DArT and DArTseq platforms. A total of 11 *Physaria* species, 1 interspecific hybrid, and 2 *Paysonia* species were analyzed by microarray DArT. Additional accessions were genotyped using the DArTseq platform comprised of 17 *Physaria* and 7 *Paysonia* species. Overall, there was a total of 177 accessions represented by single plant samples genotyped using the two platforms, majority of which are *P. fendleri* (Tables 1 and 2). The marker sequences and genotype data will be stored in the U.S. Germplasm Resources Information Network (GRIN) database (http://www.ars-grin.gov) along with the accessions' phenotypic observations and germplasm passport data curated by the U.S. National Plant Germplasm System.

Table 2. Passport information of additional Physaria and Paysonia accessions genotyped using DArTseq.

Collection No. and ID	Accession No.	Species	Country	State/Prov.	County	Latitude	Longitude	Elevation (m)

doi:10.1371/journal.pone.0064062.t002

Data Analysis

All the images from DArT platforms were analyzed using DArTsoft v.7.4.7 (DArT P/L, Canberra, Australia). The markers were scored as binary data (1/0), indicating presence or absence of a marker in genomic representation of each sample as described by Wenzl et al. [31]. For quality control, 30% of genotypes were genotyped in full technical replication. Clones with P>77%, a call rate >97% and >98% allele-calling consistency across the replicates were selected as markers. P value represents the allelic-states variance of the relative target hybridization intensity as a percentage of the total variance. The informativeness of the DArT markers was determined by calculating the polymorphism information content (PIC) within the panel of diverse accessions according to Anderson et al. [32]. The *P. fendleri* data were used in GenAlEx v.6.41[33] to determine 2D spatial autocorrelation of the DArT microarray and Arlequin v.3.5.1.3 [34] to assess the amount of variation among the assigned regional groupings by AMOVA. To summarize the relationships among all examined accessions, cluster analysis was performed on Dice similarity values with the SAHN procedure using the unweighted pair-group method done using NTSYS-pc v. 2.21 m [35]. The Dice coefficient was preferred over simple matching coefficient because DArT is a dominant marker system and there were several non-*Physaria*species in the sample set [36], [37]. In addition, a Bayesian model-based clustering was performed on microarray DArT markers using STRUCTURE v.2.3.4 [38] testing 3 independent runs with *K* from 1 to 8, each run with a burn-in period of 50,000 iterations and 300,000 Monte Carlo Markov Chain (MCMC) iterations, assuming an admixture model and correlated allele frequencies. The STRUCTURE data was subsequently analyzed by HARVESTER v.06.92 [39]. Mantel tests were made to determine if there are significant correlations between the dendrogram representations and the distance matrices, between the matrices of geographic and genetic distances, and between the distance matrices from the two DArT platforms used in the *P. fendleri*. The missing geographic coordinates of collection sites of ten accessions (PI 293027, PI 293028, PI 337050, PI 345712, PI 355037, PI 355041, PI 355042, PI 275771, PI

283700 and PI 299412) in the GRIN database were estimated using Google Earth v. 6.1.0.5001 based on available locality description and collectors' notes and the information included in the analysis. Map projections of the analyzed *P. fendleri* accessions were made using ArcGIS Explorer v.2.0.0.1750 (ESRI, Redlands, CA) and non-parametric correlation tests done using JMP v.9 (SAS Institute, Cary, NC).

RESULTS

Relationships among Accessions from Microarray DArT Analysis

A total of 2,833 polymorphic markers were found using microarray DArT, with an average genotype call rate of 98.4% and a scoring reproducibility of 99.7%. The average PIC value was 0.21 and the median 0.19. About 20% of the markers have values in the range of 0.06 to 0.10, and almost an equal proportion of 12% of markers on the following PIC classes - 0.11 to 0.15, 0.16 to 0.20, and 0.21 to 0.25 (Figure 1). Overall, the distribution of PIC values was asymmetrical and skewed towards the lower values.

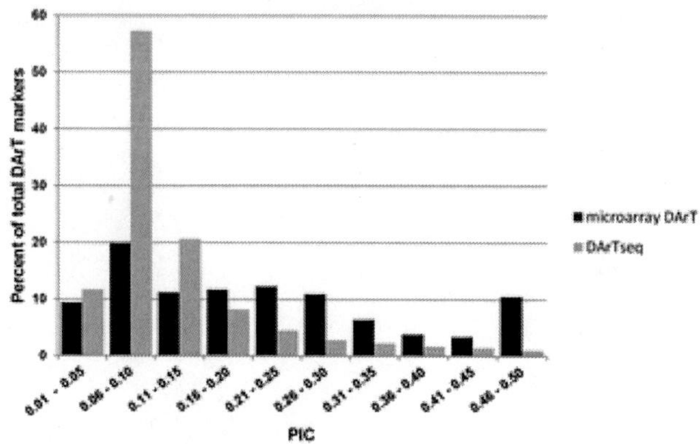

Figure 1 Distribution of marker PIC values by DArT platform.

doi:10.1371/journal.pone.0064062.g001

Cluster analysis and principal coordinate analysis (PC plot not shown) indicated that the different accessions were successfully classified by the marker system based on species, by geographical source, and breeding status (Figure 2), except for one new collection of *P. gordonii*, DDMC2010-6, which clustered with the *P. fendleri* accessions. The cophenetic correlation coefficient between the dendrogram and the distance matrix was highly significant (r = 0.98, t = 10.30, prob random Z<obs. Z = 1.00, 3000 permutations) indicating that the tree is a very good representation of the distance matrix. All *P. fendleri* accessions grouped in a separate cluster from the other species. The main cluster has two subgroups from Mexico, and a subgroup with a majority of accessions from Texas. There was no single group of *P. fendleri*accessions from Arizona and New Mexico. Most accessions from these States were associated with other accessions from Texas.

Figure 2 Cluster analysis of Physaria and Paysonia accessions based on 2,833 DArT markers.

The labels denote the germplasm collection numbers and origin. Suffixes indicate respective species (PAR: *P. argyraea,* PAU: *P. auriculata,* PFE: *P. fendleri,* PGO: *P. gordonii,* PGF: *P. grandiflora,* PGR: *P. gracilis,* PKN: *P.* 'kathryn', PLI: *P. lindheimeri,* PPL:*P. pallida,* PRC: *P. recurvata,* PRT: *P. rectipes,* and PTH: *P. thamnophila*).

doi:10.1371/journal.pone.0064062.g002

The four advanced *P. fendleri* breeding lines were partitioned into two clusters. The breeding lines WCL-SL1 and WCL-YS1 were found to be more genetically distant to WCL-LO2 and WCL-LO4. The last two lines were determined to be more similar to the rest of the *P. fendleri* from North America than WCL-SL1 and WCL-YS1.

Among the different species, the most similar to *P. fendleri* was determined to be *P. argyraea*while the least similar was *P. thamnophila*. The *P. pallida* accessions grouped in one cluster along with *P. lindheimeri* and other accessions representing the species, *P. gracilis, P. gordonii,P. recurvata,* and *P. rectipes*. The two accessions of *Paysonia auriculata* (3009) and *Paysonia grandiflora* (2243) grouped together in a separate cluster along with the interspecific hybrid swarm 'Kathryn' (4087) from five *Paysonia* species.

Theanalysisofmolecularvarianceconsidering*P.fendleri*accessions only, showed that there was a much greater proportion of variation within groups (90%) than among groups (10%) in the species (Table 3). Among the unimproved germplasm set, the pairwise Fst values that showed the greatest differentiation was between Arizona accessions and those from New Mexico and the least amount of differentiation between Mexico and Arizona (Table 4). There was a significant correlation found between the computed genetic distance and geographic distance matrices from Mantel test (r = 0.33, t = 5.94, prob random Z<obs z = 1.00, 3000 permutations), indicating that distant accession pairs are more different genetically, supporting the previously mentioned result (Figure 3).

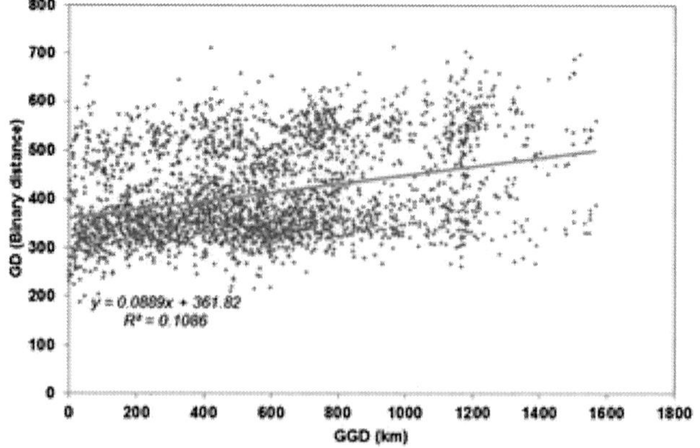

Figure 3 GenAlEx plot of geographic distance (GGD) and genetic distance (GD) from microarray DArT markers.

doi:10.1371/journal.pone.0064062.g003

Source	df	SS	Variance components	%
A. Microarray DArT				
Among Groups	4	1597.506	17.19956	9.66%
Within Groups	69	11094.561	160.79074	90.34%
Total	73	12692.068	177.99031	100%
B. DArTseq				
Among Groups	4	14299.235	98.59559	7.22%
Within Groups	123	155895.234	1267.44093	92.78%
Total	127	170194.469	1366.03651	100%

doi:10.1371/journal.pone.0064062.t003

Table 3 Analysis of molecular variance in P. fendleri accessions.

doi:10.1371/journal.pone.0064062.t003

	Arizona	Advanced lines	Mexico	New Mexico	Texas
Arizona		**0.10717**	**0.13076**	**0.03055**	**0.04049**
Advanced lines	**0.13854**		**0.22289**	**0.10951**	**0.08193**
Mexico	0.00448	**0.13043**		**0.12926**	**0.10552**
New Mexico	**0.16823**	**0.17758**	**0.16280**		**0.03056**
Texas	0.01547	**0.12599**	**0.02649**	**0.16289**	

*Values in bold are significant at $\alpha = 0.05$. Values below diagonal are from microarray DArT and those above diagonal are from DArTseq.

doi:10.1371/journal.pone.0064062.t004

Table 4 Comparison of population pairwise Fst values using microarray DArT and DArTseq.

doi:10.1371/journal.pone.0064062.t004

RELATIONSHIPS AMONG ACCESSIONS FROM DARTSEQ ANALYSIS

There was a total of 27,748 markers obtained using the *Physaria* DArTseq platform. The average genotype call rate was 98.8% and a scoring reproducibility of 99.7%. The average PIC value was 0.12 and the median was 0.09. About 57% of markers have PIC values in the range of 0.06 to 0.10, while 20% and 11% on PIC classes 0.11–0.15 and 0.01–0.05, respectively (Figure 1). The `distribution of the PIC values of DArTseq markers follows the same skewed pattern as the DArT microarray markers presented earlier.

Cluster analysis using the markers from this platform resulted in a relationship that follows that of the taxonomic groupings based on general morphological affinities presented by Rollins and Shaw [1]. The results did not deviate from those when the first platform with fewer markers used. Four accessions of *P. fendleri* (2258, 2274, 3083, and DDMC2010-8) clustered with the other*Physaria* species – *P. gordonii* and *P. gracilis*, indicating higher genetic similarity to representative accessions of these species than the rest of *P. fendleri*. These four *P. fendleri*accessions will be further examined for misidentification and for oil and morphological trait variation when verified.

The cophenetic correlation coefficient between the dendrogram and the distance matrix was highly significant (r = 0.94, t = 49.20, prob random Z<obs. Z = 1.00, 3,000 permutations) indicating that the tree is a very good representation of the distance matrix. The cluster of *P. fendleri* showed a distinct group of accessions from Mexico and a cluster comprised of all other germplasm from North America (Figure 4). The DArTseq platform indicated two separate clusters for the advanced lines. The breeding lines WCL-LH1, WCL-LO1, and WCL-LY1 are all in a group with greater similarity to accessions from Texas. WCL-LO2, WCL-LO4, WCL-SL1, and WCL-YS1 grouped together in one cluster indicating greater genetic similarity to accessions from Arizona and Mexico. WCL-LO2 was derived from the WCL-LO1 and the remaining three from WCL-LO2. It appears from these data that perhaps more genotypes from the Arizona accession were integrated into WCL-LO2.

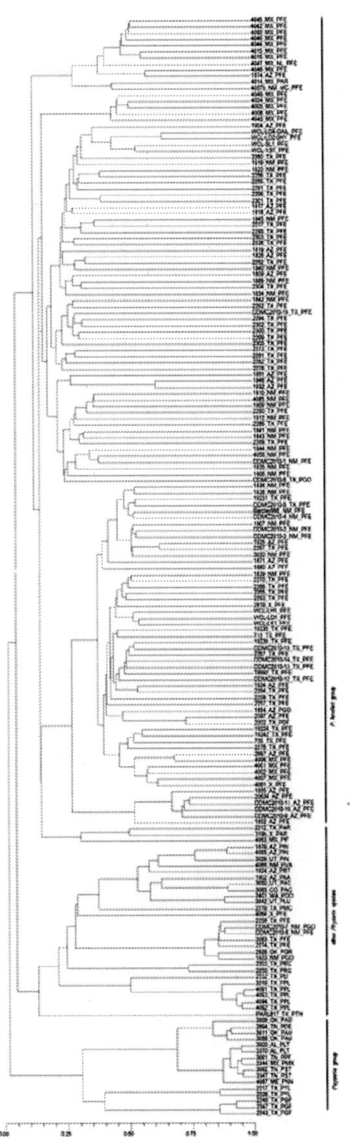

Figure 4 Cluster analysis of Physaria and Paysonia accessions based on 27,748 DArTseq markers.

The labels denote the germplasm collection numbers and origin. Suffixes indicate respective species (PAC: *P. acutifolia*, PAR: *P. argyraea*, PAU: *P. auriculata*, PDE: *P. densipila*, PDF: *P. densiflora*, PDO: *P. douglasii*, PFE: *P. fendleri*, PGO: *P. gordonii*, PGF: *P. grandiflora*,

PGR: *P. gracilis*, PIF: *P. inflata*, PIN: *P. intermedia*, PKA: *P. kaibabensis*, PKN: *P.* 'kathryn', PYL: *P. lasiocarpa*, PLI: *P. lindheimeri*, PLT: *P. lyrata*, PLU: *P. ludoviciana*, PMC: *P. mcvaughiana*, PMX: *P. mexicana*, PPF: *P. perforata*, PPL: *P. pallida*, PRC: *P. recurvata*, PRT: *P. rectipes*, PST: *P. stonensis*, PTH: *P. thamnophila* and PVA: *P. valida*).

doi:10.1371/journal.pone.0064062.g004

Similar to results from the microarray DArT, the accessions of the following species formed distinct groups consistent with the classification by Rollins and Shaw [1]: a) *P. intermedia*, *P. valida*, and *P. rectipes*, b) *P. gordonii*, *P. gracilis*, *P. rectipes*, and *P. lindheimeri*, and c) *P. auriculata*, *P. densipila*, *P. lyrata*, *P. stonensis*, *P. grandiflora*, and *P. perforata*. The two accessions of *P. lasiocarpa* (2217 and 2228) were most genetically similar to the accessions of*P. grandiflora*. This last group of species comprised of annual, auriculate-leaved types whose taxonomic nomenclature was segregated from *Physaria* and transferred to *Paysonia* based on leaf-trichome morphology, chromosome number, and molecular data from analyzing internal transcribed spacers of nuclear ribosomal DNA [40]. Apart from the accession of *Physaria mexicana* nested within the *Paysonia* cluster, the results of this genetic analysis using both microarray DArT and DArTseq platforms support the previous segregation of *Physaria* from*Paysonia* as proposed by O'Kane and Al-Shehbaz [40] indicating that the group of *Paysonia*species to be the least genetically similar to the *P. fendleri* and other *Physaria* accessions.

Results of the analysis of molecular variance when DArTseq markers were used correspond to that from microarray DArT. A greater proportion of variation within groups (93%) than among groups (7%) in the species was found (Table 3).

The average genetic similarity in the *P. fendleri* group when using microarray DArT was 0.86, while only 0.44 when DArTseq was used. This is in line with the assumption that more differences may be found when more markers are used because of the increased sensitivity and resolution to detect genetic distinctiveness [41]. The genetic similarity matrices of *P. fendleri*obtained using the two platform systems showed a good fit when compared by a Mantel test ($r = 0.48$, $t = 11.28$, prob random $Z<obs\ z = 1.00$, 3,000 permutations).

Population Structure Analysis of *P. fendleri*

Using results from microarray DArT, the population structure of the *P. fendleri* samples was determined. The plot of ΔK for each K value is shown in Figure 5a. It was estimated through the method of Evanno et. al. [42] that there are 4 groups contributing significant genetic information in the *P. fendleri* collection. The bar plot of the population assignment test when $K = 4$ is shown in Figure 5b. Three of the four accessions of *P. fendleri* breeding lines are shown to have mixed backgrounds. Of the other *P. fendleri* accessions, twelve (16%) have close to homogeneous genetic background (>98% probability) while 63 accessions (84%) are highly heterogeneous showing intermediate and/or highly mixed composition. A majority of the accessions from Texas are assigned to one cluster, while those from New Mexico have myriad cluster assignments suggesting the greater diversity of *P. fendleri* in this U.S. state. When the cluster assignments of the *P. fendleri* accessions were projected on a map, the segregation among clusters was evident in their geographic location (Figure 6). Further testing for association between the assigned clusters and available site elevation data by computing Spearman's rank correlation coefficient showed a very weak positive correlation (Spearman ρ = 0.11, p = 0.33).

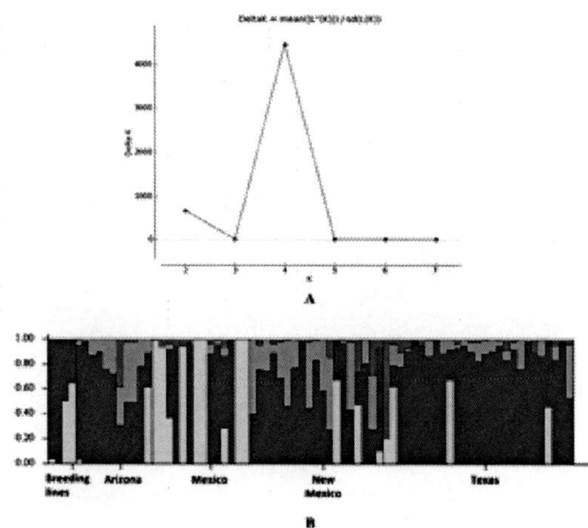

Figure 5 Plot of ΔK from K = 2 to 7 (a) and the population structure of 75 P. fendleri accessions at K = 4 (b).

doi:10.1371/journal.pone.0064062.g005

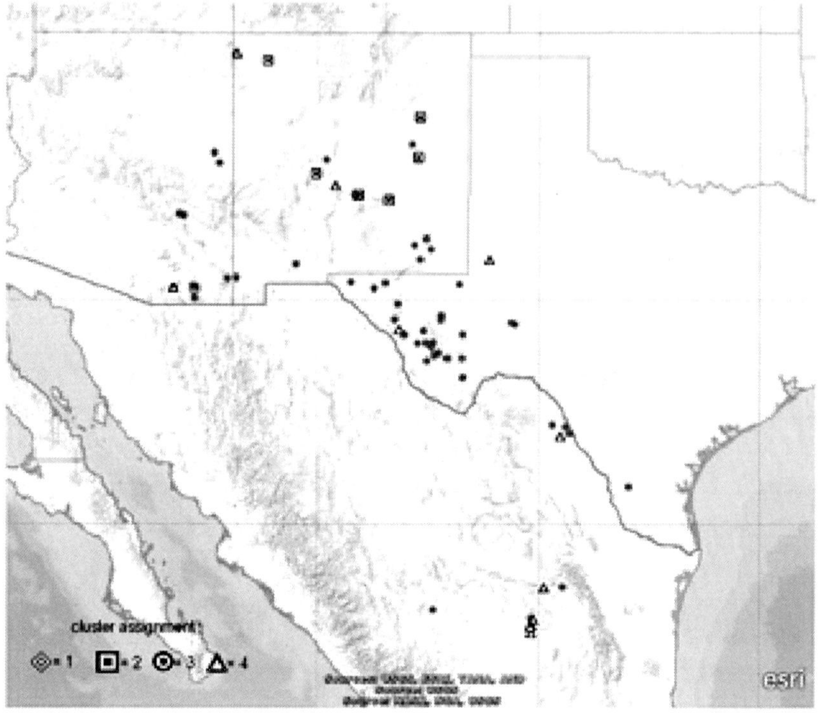

Figure 6 Geographic locations of the P. fendleri accessions and their respective cluster assignments (indicated by different icons) based on Bayesian model-based clustering methods.

doi:10.1371/journal.pone.0064062.g006

DISCUSSION

The importance of understanding genetic diversity in germplasm collections is critical for the effective management of accessions in genebanks. Molecular characterization supplements morphological evaluation of germplasm and allows measurements to help resolve numerous operational, logistical, and biological questions that face genebank managers and conservation biologists [43], [44]. The *Physaria* collection in the U.S. NPGS has not been well characterized before for genetic diversity, though there have been preliminary studies on subsets of accessions using a limited number of microsatellite markers [16], and an extensive evaluation for diversity in oil characteristics and other morphological characters [45]. In this study, the new DArT platforms for *Physaria* were found both acceptable and provided robust information about the genetic variability of the collection.

The development and utilization of DArT markers allowed us to determine the genetic diversity of the *Physaria* collection. The 2,833 microarray DArT markers were found to be useful in providing a picture of genetic diversity in the *Physaria* germplasm collection using a large set of accessions. Overall, the average PIC of the *Physaria* and *Paysonia* microarray DArT markers was found to be lower than that observed in other species where similar markers were developed, like wheat (0.44) [46], cassava (0.42) [47], and sorghum (0.41) [48], but comparable to that observed in sugar beet (0.28) [49] and *Asplenium* fern (0.21) [50]. The average PIC of the DArTseq markers is much less than that of the microarray DArT. However, the more numerous DArTseq markers may have the capability of providing a better picture of diversity by sampling more points in the genome. The distribution of these almost 30,000 new DArT markers in the *Physaria* and *Paysonia* genome remains to be determined. However, based on the information from a large number of organisms in which DArT system was applied more broadly including genetic mapping and/or sequence-based physical mapping, we can assume that DArT marker from both platforms will be distributed throughout the genome with marker density highly correlated to gene density [28], [51]. Compared to microsatellite markers, DArT markers are very suitable for high-throughput work and previously have been determined to have clear advantages in cost and time aspects of genotyping as demonstrated in other crops [52]. Both microarray DArT and DArTseq platforms have the same development costs. However, the higher number of markers obtained in DArTseq resulted to an overall lower cost per datapoint than microarray DArT. This higher cost effectiveness of DArTseq is in parallel to other sequenced-based genotyping strategies which can provide substantial cost savings compared to microarrays when conducting genetic diversity studies [53], [54]. Importantly, when comparing*effectiveness* of the two platforms one has to keep in mind that it may vary according to specific*application*: in genetic ID and product quality testing (i.e. seed purity) modest number of array-based DArT markers may perform as well as DArTseq platform and currently for a better price. A further cost reduction of sequencing may however push even this balance towards DArTseq platform in the future.

The relationships found among the accessions are in line with the previously proposed evolution within the genus. Taxonomists assert that *P. auriculata* is the most primitive species due to its very distinct evolutionarily primitive characters such as large siliques, large number of ovules around the replum, and predominance of simple trichomes [1]. *P. auriculata* has been proposed to be closely related to *P. grandiflora* and this relationship is supported by the results of the molecular marker analysis between the accessions representative of these species (2243 and 3009) showing high genetic similarities. Likewise, *P. gordonii* and *P. gracilis* grouped in the same cluster with *P. rectipes, P. recurvata,* and *P. thamnophila* which is in agreement with their previously known taxonomic groupings based on very general morphological similarities. The high genetic similarity between the *P. argyraea* accessions (2212 and 319 b) to the *P. fendleri*group supports phylogenetic subsectional grouping based on pod morphology.

Overall, there was higher genetic similarity found among accessions of the other *Physaria* and*Paysonia* species than among accessions of *P. fendleri*. In particular, species that are included in the federal or state threatened and endangered species list, like *P. pallida* and *P. stonensis*[55], have very low genetic diversity as indicated by results of DArTseq markers. In *P. pallida,* the representative accessions (4091 and 4093) were found to be highly similar. The limited geographic range and the proximity of the collection sites of these two samples suggest that they might have been from just one population. Likewise for *P. stonensis,* there was a very high genetic similarity found on both of the representative accessions (3092 and 3347). Because only a limited number of accessions were included in these other species, a follow up study that includes additional accessions is recommended to validate if this is a general trend.

Physaria 'Kathryn' (4087) is a cultivar from interspecific hybridization developed by allowing five species (*P. densipila,*

P. lescurii, P. lyrata, P. perforata, and *P. stonensis*) to intermate for twenty two generations [56]. *P.* 'Kathryn' was determined to be most closely related to *P. auriculata* in our analysis using microarray DArT. However, with the expanded set of accessions included in the DArTseq analysis, it grouped with the representative samples of its parent species - *P. lyrata* (3000 and 3370), *P. perforata* (3091), and *P. stonensis* (3092, 3347). These species were also found most genetically similar to *P. mexicana* (3344) which is the only perennial type in the species cluster. *P. mexicana* is a previously undescribed species in Mexico and is among the more recent species reported by Rollins [57].

DDMC2010-6 was from a more recent germplasm collecting trip and it was entered in the database as *P. gordonii*. However, based on results of DArT markers, it clustered with the *P. fendleri* accessions after using both DArT platforms, indicating the need to review its species assignment. The species identity of this particular accession will again be verified using its plant voucher specimen as well as in the NPGS site handling the germplasm when it is regenerated in the future.

The two clusters of *P. fendleri* breeding lines - WCL-LO2, WCL-LO4, WCL-SL1, and WCL-YS1 in one cluster, while the other set of breeding lines WCL-LO1, WCL-LH1, and WCL-LY1 in another, indicates the possibility of developing genetically differentiated lines for crop improvement and may have applications in future hybrid development work. The lines WCL-LH1, WCL-LO1 and WCL-LY1 are the first three germplasm lines that were publicly released in 1996. These were developed using recurrent selection on a population made by bulking seeds of one accession that came from Arizona and nine from Texas in 1986 [58]. The bulked seeds were also the starting material for other breeding lines. WCL-LO4 was derived from mass selection from WCL-LO3 and WCL-LY2 (both not included in this study) which has WCL-LY1 as the source population [59], [60]. The other breeding lines were developed through phenotypic selection: WCL-YS1 was a selection from PI 311165, one of the initial accession from Arizona that comprise the original bulk in 1986 [61], while WCL-SL1 came from plants that survived at the highest salinity levels during a salt tolerance screening study when seeds from the original bulked seeds were planted [62].

The accessions of *P. fendleri* from Mexico are genetically similar as indicated by the cluster analyses. The range of *Physaria* species has been reported as limited to the northeastern part of country and concentrated on mountain and high plains of Coahuila, Nuevo León, and Zacatecas [57]. This limited geographic distribution likely prevented their further genetic differentiation. A similar investigation focusing on the other *Physaria* species in this region could confirm this trend.

The array of *P. fendleri* germplasm consisted of 75 accessions analyzed using microarray DArT and 128 accessions using DArTseq. There is ample genetic variability in the *P. fendleri*collection found, as indicated by the cluster analysis as well as analysis of population structure. This is expected for a cross pollinating species that has not been fully domesticated [63]. However, the higher within group variation detected by AMOVA using data from both DArT platforms suggests that there is only a small amount of genetic differentiation among groups in the sample as a whole. Results from the Bayesian clustering approach when comparing the geographical sources of the accessions suggested that there is more variability in New Mexico than the other *P. fendleri* locations and that there is a spatial pattern evident in the microarray DArT results. This pattern of genetic differentiation occurs outward from central Texas, the region identified as the putative origin of *Physaria* as proposed earlier by Payson [64].

An increased population differentiation has been reported in many plant species between source populations and new ones when plants colonize new habitats [65]. The more dynamic nature of *Physaria* populations in distant locations from Texas has been reported by Payson[64] and he attributed part of the process as caused by barriers that separate populations, such as soil properties and moisture availability which are very important to survival of ephemeral populations of the taxon. In *P. fendleri,* Dierig et al. [66] stated that temperature and elevation effects can also account for significant differences in reproductive capacity. Selection patterns have also been investigated by past studies reporting that non-random mating, sexual selection and dormancy characteristics play a significant role in how traits evolved in the species [67],[68]. Seed dormancy characteristics in particular cause a persistent soil seed bank which may prevent genetic differentiation in the species because certain genotypes are reintroduced back during subsequent seasons [69], [70]. In terms of genetic resources conservation of *P. fendleri*germplasm, the DArT results suggest the existence of more variable genotypes in New Mexico and it is recommended that future collecting missions select this geographic area to possibly expand the genetic diversity in the species collection.

CONCLUSION

The availability of genetic diversity information of the *P. fendleri* collection will enable better germplasm management and conservation of the species. In this study we report the successful development of two DArT marker platforms that were utilized for genotyping *Physaria* and*Paysonia* accessions. This marker system complements the microsatellite markers developed previously for *Physaria*. The high number of DArT markers allows a greater resolution of genetic differences among accessions and enabled us to examine the extent of variation in the *P. fendleri* collection, as well as provide support to known taxonomic classification and recent nomenclatural changes of certain *Physaria* species to *Paysonia*. We intend to further utilize the DArT markers in developing a linkage map in *Physaria* to assist breeding efforts and for future genetic mapping studies.

ACKNOWLEDGMENTS

USDA is an equal opportunity provider and employer. Mention of companies or commercial products does not imply recommendation or endorsement by the USDA over others not mentioned. USDA neither guarantees nor warrants the standard of any product mentioned. Product names are mentioned solely to report factually on available data and to provide specific information. We would like to thank David Ellis and Chris Richards for the valuable suggestions during the conduct of the study.

AUTHOR CONTRIBUTIONS

Conceived and designed the experiments: VMVC AK DAD.

Performed the experiments: VMVC AK DAD. Analyzed the data: VMVC AK. Wrote the paper: VMVC AK DAD. Participated in germplasm collection: VMVC DAD.

REFERENCES

1. Rollins RC, Shaw EA (1973) The Genus *Lesquerella* (Cruciferae) in North America. Harvard Univ. Press, Cambridge, MA. 288 p.
2. Al-Shehbaz IA, O'Kane SL Jr (2002) *Lesquerella* is united with *Physaria* (Brassicaceae). Novon 12: 319–329. doi: 10.2307/3393073
3. Kish S (2008) *Lesquerella*: The next source of biofuel. USDA Cooperative State Research, Education, and Extension Service (CSREES). Available:http://www.csrees.usda.gov/newsroom/impact/2008/nri/pdf/lequerella.pdf. Accessed 2012 March 29.
4. Moser BR, Cermak SC, Isbell TA (2008) Evaluation of castor and lesquerella oil derivatives as additives in biodiesel and ultralow sulfur diesel fuels. Energy & Fuels 22: 1349–1352. doi: 10.2307/3393073
5. Dierig DA, Ray DT (2009) New Crops Breeding: *Lesquerella.* In: Vollman J, Rajcan I (eds). Oil Crops, Handbook of Plant Breeding 4. Springer Science Media, NY. 507–516.
6. Cruz VMV, Dierig DA (2012) Trends in literature on new oilseed crops and related species: Seeking evidence of increasing or waning interest. Industrial Crops and Products 37: 141–148.
7. Jenderek MM, Dierig DA, Isbell TA (2009) Fatty-acid profile of *Lesquerella* germplasm in the National Plant Germplasm System collection. Industrial Crops and Products 29: 154–164.
8. Wang GS, McCloskey W, Foster M, Dierig D (2010) Lesquerella: A winter oilseed crop for the Southwest. Arizona Cooperative Extension. The University of Arizona, Tucson, AZ. 4 p.
9. Cruz VMV, Dierig DA (2010) National conservation activities on *Lesquerella* spp. for ensuring species survival and germplasm availability for new crops research and development. Proc Colorado Native Plant Society Annual Meeting. Sept. 10–12, 2010. Denver, CO.
10. Mondini L, Noorani A, Pagnotta MA (2009) Assessing plant genetic diversity by molecular tools. Diversity 1: 19–35.
11. FAO (2010) The Second Report on the State of the World's Plant Genetic Resources for Food and Agriculture. Commission on Genetic Resources for Food and Agriculture. Food and Agriculture Organization of the United Nations. Rome, Italy. 69–70.

12. Börner A, Khlestkina EK, Pshenichnikova TA, Osipova SV, Kobiljski B, et al.. (2012) Genetics and genomics of plant genetic resources. Journal of Stress Physiology & Biochemistry 8 suppl. vol. 3, S10.

13. Cruz VMV, Nason J, Luhman R, Marek LF, Shoemaker RC, et al. (2006) Analysis of bulked and redundant accessions of *Brassica* germplasm using assignment tests of microsatellite markers. Euphytica 152: 339–349. doi: 10.1007/s10681-006-9221-5

14. Cruz VMV, Luhman R, Marek LF, Rife CL, Shoemaker RC, et al. (2007) Characterization of flowering time and SSR marker analysis of spring and winter type *Brassica napus* L. germplasm. Euphytica 153: 43–57. doi: 10.1007/s10681-006-9221-5

15. Hasan M, Seyis F, Badani AG, Pons-Kühnemann J, Friedt W, et al. (2006) Analysis of genetic diversity in the *Brassica napus* L. gene pool using SSR markers. Genetic Resources and Crop Evolution 53: 793–802. doi: 10.1007/s10681-006-9221-5

16. Salywon AM, Dierig DA (2006) Isolation and characterization of microsatellite loci in*Lesquerella fendleri* (Brassicaceae) and cross-species amplification. Mol Ecol Notes 6: 382–384.

17. Redden R, Vardy M, Edwards D, Raman H, Batley J (2009) Genetic and morphological diversity in the Brassicas and wild relatives. Proc. 16th Australian Research Assembly on Brassicas. Sept. 14–16, 2009. Ballarat, Victoria. 5 p.

18. Faltusová Z, Kučera L, Ovesná J (2011) Genetic diversity of *Brassica oleracea* var.*capitata* Gene Bank accessions assessed by AFLP. Electronic Journal of Biotechnology.http://dx.doi.org/10.2225/vol14-issue3-fulltext-4http://dx.doi.org/10.2225/vol14-issue3-fulltext-4. Accessed 2012 March 29.

19. Toledo J, Dehal P, Jarrin F, Hu J, Hermann M, et al. (1998) Genetic variability of*Lepidium meyenii* and other Andean *Lepidium* species (Brassicaceae) assessed by molecular markers. Ann Bot 82: 523–530.

20. Kamel EA, Hassan HZ, El-Nahas AI, Ahmed SM (2004) Molecular characterization of some taxa of the genus *Raphanus* L. (Cruciferae = Brassicaceae). Cytologia 69: 249–260

21. Jaccoud D, Peng K, Feinstein D, Kilian A (2001) Diversity Arrays: a solid state technology for sequence information independent genotyping. Nucleic Acids Res 29: e25. doi: 10.1093/nar/29.4.e25

22. Cruz, unpublished data.

23. Jones N, Ougham H, Thomas H, Pasakinskiene I (2009) Markers and mapping revisited: finding your gene. New Phytologist 137: 165–177. doi: 10.1111/j.1469-8137.2009.02933.x

24. DArT (2012) Papers about DArt. Diversity Arrays Technology, Pty Ltd. Available:http://www.diversityarrays.com/publications.html. Accessed 2012 March 30.

25. Varshney RK, Glaszmann JC, Leung H, Ribaut JM (2010) More genomic resources for less-studied crops. Trends in Biotechnology 28: 452–460. doi: 10.1016/j.tibtech.2010.06.007

26. Howard EL, Whittock SP, Jakše J, Carling J, Matthews PD, et al. (2011) High-throughput genotyping of hop (*Humulus lupulus* L.) utilising diversity arrays technology (DArT). Theor Appl Genet 122: 1265–1280. doi: 10.1007/s00122-011-1529-4

27. Cabin RJ, Mitchell RJ, Marshall DL (1998) Do surface plant and soil seed bank populations differ genetically? A multipopulation study of the desert mustard *Lesquerella fendleri* (Brassicaceae). Am J Bot 85: 1098–1109. doi: 10.2307/2446343

28. Kilian A, Wenzl P, Huttner E, Carling J, Xia L, et al. (2012) Diversity Arrays Technology: A Generic Genome Profiling Technology on Open Platforms. Methods Mol Bio 888: 67–88. doi: 10.1007/978-1-61779-870-2_5

29. O'Kane SL Jr, Al-Shehbaz IA (2003) Phylogenetic position and generic limits of*Arabidopsis* (Brassicaceae) based on sequences of nuclear ribosomal DNA. Ann. Missouri Bot. Gard. 90: 603–612. doi: 10.2307/3298545

30. Langmead B, Trapnell C, Pop M, Salzberg SL (2009) Ultrafast and memory-efficient alignment of short DNA sequences to the human genome. Genome Biol 10: R25. doi: 10.1186/gb-2009-10-3-r25

31. Wenzl P, Carling J, Kudrna D, Jaccoud D, Huttner E, et al. (2004) Diversity arrays technology (DArT) for whole genome profiling of barley. Proc Natl Acad Sci (USA) 101: 9915–9920. doi: 10.1073/pnas.0401076101

32. Anderson JA, Churchill GA, Sutrique JE, Tanksley SD, Sorrells ME (1993) Optimizing parental selection for genetic linkage maps. Genome 36: 181–186. doi: 10.1139/g93-024

33. Peakall R, Smouse P (2006) GenalEx 6: Genetic Analysis in Excel. Population genetic software for teaching and research. Mol Ecol Notes 6, 288–295.

34. Excoffier L, Laval G, Schneider S (2005) Arlequin ver. 3.0: An integrated software package for population genetics data analysis. Evolutionary Bioinformatics Online 1: 47–50.

35. Rohlf FJ (2011) NTSYSpc: Numerical Taxonomy System, ver. 2.21 m. Exeter Publishing, Ltd., Setauket, NY.

36. Dalirsefat S, Meyer A, Mirhoseini S (2009) Comparison of similarity coefficients used for cluster analysis with amplified fragment length polymorphism markers in the silkworm,*Bombyx mori*. Journal of Insect Science 9: 71. doi: 10.1673/031.009.7101

37. Sesli M, Yegenoglu ED (2010) Comparison of similarity coefficients used for cluster analysis based on RAPD markers in wild olives. Genet Mol Res 9: 2248–2253. doi: 10.4238/vol9-4gmr966

38. Pritchard JK, Stephens M, Donnelly P (2000) Inference of population structure using multilocus genotype data. Genetics 155: 945–959. doi: 10.1673/031.009.7101

39. Earl DA, vonHoldt BM (2012) STRUCTURE HARVESTER: a website and program for visualizing STRUCTURE output and implementing the Evanno method. Conservation Genetics Resources 4: 359–361. doi: 10.1673/031.009.7101

40. O'Kane SL Jr, Al-Shehbaz IA (2002) *Paysonia*, a new genus segregated from*Lesquerella* (Brassicaceae). Novon 12: 379–381. doi: 10.1673/031.009.7101

41. Agarwal M, Shrivastava N, Padh H (2008) Advances in molecular marker techniques and their applications in plant sciences. Plant Cell Rep 27: 617–631. doi: 10.1007/s00299-008-0507-z

42. Evanno G, Regnaut S, Goudet J (2005) Detecting the number of clusters of individuals using the software STRUCTURE: a simulation study. Mol Ecol 14: 2611–2620. doi: 10.1111/j.1365-294X.2005.02553.x

43. Hedrick PW (2000) Applications of population genetics and molecular techniques to conservation biology. In: Conservation Biology 4. Genetics, Demography and Viability of Fragmented Populations. Young, A.G, Clarke, G.M. (eds.). Cambridge Univ. Press, Cambridge, UK. 113–125.

44. Karp A, Kresovich S, Bhat KV, Ayad WG, Hodgkin T (1997) Molecular tools in plant genetic resources conservation: a guide to the technologies. IPGRI Technical Bulletin No. 2. International Plant Genetic Resources Institute, Rome, Italy. 47 p.

45. Salywon AM, Dierig DA, Rebman JP, Jasso de Rodriguez D (2005) Evaluation of new*Lesquerella* and *Physaria* (Brassicaceae) oilseed germplasm. Am J Bot 92: 53–62. doi: 10.3732/ajb.92.1.53

46. Raman H, Stodart BJ, Cavanagh C, Mackay M, Morell M, et al. (2010) Molecular diversity and genetic structure of modern and traditional landrace cultivars of wheat (*Triticum aestivum* L.). Crop and Pasture Science 61: 222–229.

47. Xia L, Peng KM, Yang SY, Wenzl P, de Vicente MC, et al. (2005) DArT

for high-throughput genotyping of cassava (*Manihot esculenta*) and its wild relatives. Theor Applied Genet 110: 1092–1098. doi: 10.1007/s00122-005-1937-4

48. Mace ES, Xia L, Jordan DR, Halloran K, Path DK, et al.. (2008) DArT markers: diversity analyses and mapping in *Sorghum bicolor*. BMC Genomics. 9, 26. doi: 10.1186/1471-2164-9-26. Accessed 2012 March 29.
49. Simko I, Eujayl I, van Hintum TJL (2012) Empirical evaluation of DArT, SNP, and SSR marker-systems for genotyping, clustering, and assigning sugar beet hybrid varieties into populations. Plant Science 184: 54–62. doi: 10.1016/j.plantsci.2011.12.009
50. James KE, Schneider H, Ansell SW, Evers M, Robba L, et al. (2008) Diversity Arrays Technology (DArT) for pan-genomic evolutionary studies of non-model organisms. PLoS ONE 3: e1682. doi: 10.1371/journal.pone.0001682
51. Petroli CD, Sansaloni CP, Carling J, Steane DA, Vaillancourt RE, et al. (2012) Genomic characterization of DArT markers based on high-density linkage analysis and physical mapping to the *Eucalyptus* genome. PLoS ONE 7: e44684. doi: 10.1371/journal.pone.0044684
52. Kilian A, Huttner E, Wenzl P, Jaccoud D, Carling J, et al.. (2005) The fast and the cheap: SNP and DArT-based whole genome profiling for crop improvement. In: Tuberosa R, Phillips RL, Gale M (eds) Proceedings of the international congress in the wake of the double helix: from the green revolution to the gene revolution. Avenue Media, Bologna, Italy, 27–31 May 2003, 443–461.
53. Illumina (2013) Agrigenomics genotyping decisions reach a crossroads. Application Spotlight: Analyzing Genetic Variation. Available:http://www.illumina.com/Documents/products/appspotlights/app_spotlight_ngg_ag.pdf. Accessed 2013 February 13.
54. Ayling S (2012) Technical appraisal of strategic approaches to large-scale germplasm evaluation. The Genome Analysis Center, Norwich, UK. Available:http://agro.biodiver.se/wp-content/uploads/2012/12/Technical-appraisal-NGS-for-genebanks-please-comment.pdf. Accessed 2013 February 13.
55. USDA-NCRS (2012) The PLANTS Database. National Plant Data Team, Greensboro, NC. Available: http://plants.usda.gov. Accessed 2012 September 24.
56. Rollins RC (1988) A population of interspecific hybrids of *Lesquerella* (Cruciferae). Systematic Botany 13: 60–63.

57. Rollins RC (1958) Notes on *Lesquerella* (Cruciferae) in Mexico. Boletin de la Sociedad Botanica de Mexico 23: 42–47.
58. Dierig DA, Thompson AE, Coffelt TA (1998) Registration of three *Lesquerella fendleri*germplasm lines selected for improved oil traits. Crop Sci 38: 287.
59. Dierig DA, Dahlquist GH, Coffelt TA, Ray DT, Isbell TA, et al. Registration of WCL-LO4-Gail lesquerella with improved harvest index. J Plant Reg "submitted".
60. Dierig DA, Dahlquist GA, Tomasi PM (2006b) Registration of WCL-LO3 high oil*Lesquerella fendleri* germplasm. Crop Sci 46: 1832–1833.
61. Dierig DA, Tomasi PM, Coffelt TA, Rayford WE, Lauver L (2000) Yellow seed coat*Lesquerella*. Crop Sci 40: 865.
62. Dierig DA, Shannon MC, Grieve CM (2001) Salt tolerant *Lesquerella*. Crop Sci 41: 604.
63. Rauf S, Teixeira da Silva JA, Khan AA, Naveed A (2010) Consequences of plant breeding on genetic diversity. Intl J Plant Breeding 4: 1–21.
64. Payson EB (1921) A Monograph of the Genus *Lesquerella*. Annals of the Missouri Botanical Garden 8: 103–236.
65. Barrett SCH, Husband BC (1990) The Genetics of Plant Migration and Colonization. In: Brown AHD, Clegg MT, Kahler AL, Weir BS (eds). Plant population genetics, breeding, and genetic resources. Sinauer Associates Inc., Sunderland, MA. 254–277.
66. Dierig DA, Adam NR, Mackey NR, Dahlquist GH, Coffelt TA (2006a) Temperature and elevation effects on plant growth, development, and seed production of two *Lesquerella*species. Industrial Crops and Products 24: 17–25. doi: 10.1016/j.indcrop.2005.10.004
67. Cabin RJ, Evans AS, Mitchell RJ (1997) Do plants derived from seeds that readily germinate differ from plants derived from seeds that require forcing to germinate? A case study of the desert mustard *Lesquerella fendleri*. Ann Midl Nat 138: 121–133. doi: 10.1016/j.indcrop.2005.10.004
68. Mitchell RJ, Marshall DL (1998) Nonrandom mating and sexual selection in a desert mustard: an experimental approach. Am J Bot 85: 48–55. doi: 10.2307/2446553
69. Cabin RJ (1996) Genetic comparisons of seed bank and seedling populations of a perennial desert mustard, *Lesquerella fendleri*. Evolution 50: 1830–1841. doi: 10.1016/j.indcrop.2005.10.004
70. Freeland J (2005) Molecular Ecology. John Wiley & Sons, Ltd., Hoboken, NJ. 109–154.

Chapter 2

BREEDING BASED REMOBILIZATION OF TOL2 TRANSPOSON IN XENOPUS TROPICALIS

Maura A. Lane, Megan Kimber, Mustafa K. Khokha

Program in Vertebrate Developmental Biology, Department of Pediatrics and Genetics, Yale University School of Medicine, New Haven, Connecticut, United States of America

ABSTRACT

Xenopus is a powerful model for studying a diverse array of biological processes. However, despite multiple methods for transgenesis, relatively few transgenic reporter lines are available and commonly used. Previous work has demonstrated that transposon based strategies are effective for generating transgenic lines in both invertebrate and vertebrate systems. Here we show that the Tol2 transposon can be remobilized in the genome of X. tropicalis and passed through the germline via a simple breeding strategy of crossing transposase expressing and transposon lines. This remobilization system provides another tool to exploit transgenesis and opens new opportunities for gene trap and enhancer trap strategies.

INTRODUCTION

The frog, Xenopus, is an ideal vertebrate model for human biology given its evolutionary position among the vertebrates, low cost, and large quantity of embryos amenable to manipulation and molecular techniques [1,2]. To fully exploit Xenopus, transgenic reporter lines are necessary to visualize subcellular structures, specific tissues, or dynamic cellular processes [3-6]. Despite the many available transgenesis tools, however, few reporter lines exist in X. tropicalis.

An effective scheme for generating reporter lines is enhancer or gene trapping, whereby a transgenic cassette is randomly inserted into the genome and co-opts the expression pattern of a nearby gene [7-14]. Transposon systems are commonly used to introduce these cassettes. One such system is based on Tol2, a member of the hAT (hobo, Ac, and Tam) DNA transposon family, originally isolated as an autonomous element from the Medaka fish Orzyias latipes[15,16], and developed as a non-autonomous system [17] for use in transgenesis. This "cut and paste" transposon inserts canonically at a random heterogenic sequence, often at multiple loci, and creates a signature eight base pair (bp) target site duplication (TSD) [18]. Tol2 has applications for Xenopus transgenesis [19-21], has demonstrated "local hopping" [22-25], and has been shown to favor insertion into transcriptional regions of genes [24-26], all properties which make it highly attractive as a gene and enhancer trapping tool in frog.

Sleeping Beauty, in the Tc1/mariner family of transposons, is another transposon system commonly used to insert transgenic cassettes into the genome. It differs from Tol2 in a number of characteristics affecting integration or remobilization of transposons. For example, Sleeping Beauty targets a random TA sequence for integration, shows a preference to integrate a low copy number concatomer in intergenic regions, and has a very high tendency for local hopping [26,27].

Once inserted into the genome, transposons can be remobilized by transient transposase re-expression, resulting in novel integrations and further opportunities for enhancer trapping. In Drosophila and

Zebrafish, transposon remobilization and screening of transgenic lines is common and effective [23,24,28-30]. In X. tropicalis, remobilization of transposons has been demonstrated using two methods: 1) microinjection of transposase mRNA (Tol2) [22] and 2) breeding transposon lines to transposase lines (Sleeping Beauty) [27].

In the microinjection method [22], *in vitro* transcribed Tol2 mRNA injected in embryos bearing the Tol2 transposon resulted in a very high remobilization efficiency (18/18 injected embryos showed evidence of remobilization) [22]. The micro-injection method is labor intensive however, as transposase mRNA is prepared and embryo injections performed each time remobilization is desired.

In the breeding based remobilization strategy employing Sleeping Beauty [27], a transgenic animal with the Sleeping Beauty transposon was crossed with one from another line harboring a corresponding transposase expressing construct. Exposure to the transposase resulted in remobilization and re-integration of substrate transposons. This breeding based strategy has thus far proven less efficient than mRNA injection in producing remobilization and reintegration in Xenopus, less than 1% on average [27]. While less efficient than mRNA injection, the breeding based strategy is less labor intensive, only requiring setting up a mating, then collecting and analyzing embryos.

As with injection based transposon transgenesis methods, it is advantageous to have multiple types of transposon systems available for use with breeding based remobilization strategies. Thus, a transposon system such as Sleeping Beauty may be useful in gene therapy applications, where non-genic insertions are desirable, while one such as Tol2 may be useful for enhancer trap screens, where genic insertions are valuable [25]. Multiple transposon systems used together could also harness the advantages of each for different applications such as enhancer or gene trapping.

Currently, *Xenopus* lacks the remarkable repertoire of transgenic lines available for other systems. A breeding based remobilization strategy using Tol2 would provide a new platform to create currently lacking reporter lines by enhancer or gene trapping in *Xenopus*. Here we demonstrate this strategy for the first time by crossing Tol2 transposon and transposase lines inX. *tropicalis*. We identify readily apparent remobilization in an average of 1.3% of embryos, and we are able to successfully map new integrations in 67% of these embryos.

MATERIALS AND METHODS

Ethics Statement

The animal protocol was approved by the Yale University Institutional Animal Care and Use Committee (Animal Use Protocol Number: 2012-11035).

Husbandry

X. tropicalis were housed and cared for in our aquatics facility according to established protocols approved by the Yale University Institutional Animal Care and Use Committee.

Plasmids and generation of transgenic lines

Tol2 transposon *X. tropicalis* animals.

We obtained heterozygous Tol2 transposon animals derived from the "10M" female founder, who harbors a single Tol 2 insertion mapping to scaffold 8 of the JGI Genome v. 7.1 (animals kindly provided by Paul Mead) [20]. This transposon, which we call Ef1αGFPTol2, contains an EF1α enhancer which drives eGFP ubiquitously in the embryo (Figure 1C). We confirmed the insertion locus in our heterozygous animals by linker-mediated PCR (LM-PCR, see below). This insertion site became our substrate for testing transposon remobilization. We bred the heterozygous animals to homozygosity, and selected three homozygous animals as F1 founders to cross with our F1 transposase animals (described below and Figure 2 and Figure S1).

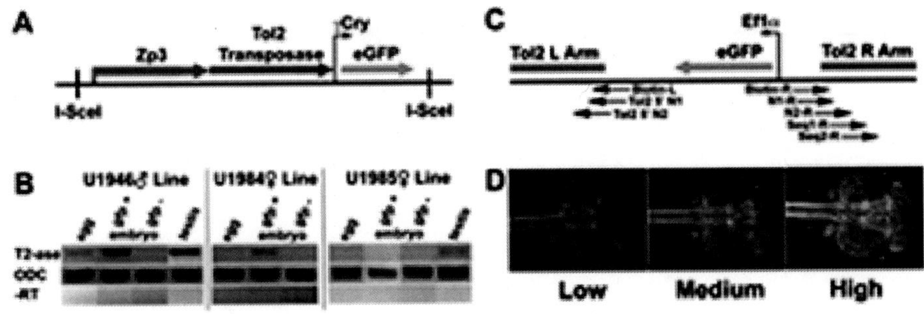

Figure 1. Transgenic lines. A. Diagram of ZP3T2γGMN construct used to produce transgenic frogs expressing Tol2 transposase. Not to scale. I-SceI: Meganuclease site necessary for transgenesis. Zebrafish zona pellucida glycoprotein 3 (zp3) promoter: drives egg specific expression of Tol2 transposase. Gamma crystallin promoter (γCry): drives expression of eGFP in lens of the eye as a reporter for transgene insertion. B. RT-PCR for transposase expression in transgenic ZP3T2γGMN F1 offspring arising from outcrosses of transposase transgene injected animals U1946♂, U1984♀ and U1985♀. Pools of 10 egg, stage 30 gfp+ and gfp- embryos were tested. In the U1946♂ and U1985♀ lines, one testis from adult male frogs was also tested. OCD primers were used as positive controls (- RT). reactions using GFP+ embryos, and water (data not shown) were also used as negative controls. C. Diagram of Ef1αGFPTol2 construct. Tol2 left (L) and right (R) arms, EF1α enhancer driving eGFP transgene. Arrows: indicate specified LM-PCR and sequencing primer binding sites on transposon arms. D. Ubiquitous GFP+ phenotypes. Low, Medium, and High Intensity phenotypes were seen in F3 and F4 embryos.

doi:10.1371/journal.pone.0076807.g001

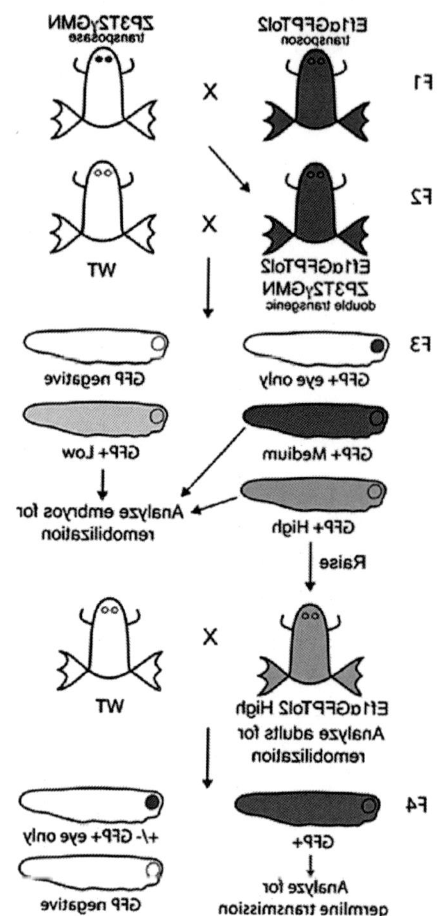

Figure 2. Mating scheme for generation and analysis of Ef1αGFPTol2 remobilized transposons. The diagram is a schematic of the crosses to test transposon mobilization in each of three different lines. We crossed three F1 heterozygous ZP3T2γMN transposase frogs (one from each injected tranposase animal U1946♂, U1984♀ and U1985♀) with three F1 homozygous Ef1αGFPTol2 frogs to generate three F2 double transgenic (Ef1αGFPTol2/ZP3T2γGMN) offspring clutches. Double transgenic animals from each clutch were outcrossed, and F3 ubiquitous GFP+ embryos of each phenotype observed were collected and tested for remobilization. High intensity F3 embryos, found only in the U1984♀ line, were raised to adulthood, tested for remobilization, and outcrossed to test for germline transmission.

doi:10.1371/journal.pone.0076807.g002

GENERATION OF ZP3T2ΓGMN TRANSPOSASE *X. TROPICALIS* LINES.

We cloned the zp3 (zona pellucida 3) promoter (kindly provided by Z. Gong) [31] adjacent to the Tol2 transposase (T2; kindly provided by K. Kawakami) [32]. Adjacent to this cassette, we included a γCrystallin-eGFP cassette in order to follow transposase inheritance in our frogs (γG; kindly provided by Rob Grainger) [3]. This ZP3T2γG cassette was then subcloned into a meganuclease vector with flanking I-SceI sites (Figure 1A).

We created three ZP3T2γGMN transposase transgenic lines using the meganuclease method [33], which tends to result in a concatomer of multiple transgene integrations at a single locus. Three injected ZP3T2γGMN animals (U1946♂, U1984♀, and U1985♀, Figure S1) showing GFP expression in the eye (GFP+eye) at stages 38-40 [34] were raised and outcrossed with WT frogs to create three F1 clutches with GFP+ eye expression (C882, C913, C914 respectively, Figure S1). In order to identify F1 animals bearing a single insertion locus of the transposase transgene, a subset of F1 animals from each clutch were raised and outcrossed to wild-type animals (Figure S1). We selected three F1 animals with transgene transmission rates consistent with a single insertion (i.e 50% of F2 animals were GFP+eye) to cross with our Ef1αGFPTol2 transposon animals (Figure 2 and Figure S1). Anticipating that different expression levels of transposase may lead to different transposon remobilization efficiencies, we did not characterize the insertion locus for copy number by Southern blot, but instead chose to analyze expression levels by RT-PCR, and then empirically test for remobilization rates. For simplicity, we call the three transposase lines the U1946♂, U1984♀, and U1985♀ lines, referring to the original injected animal giving rise to each line (Figure S1).

GENERATION OF DOUBLE TRANSGENIC EF1 GFPTOL2 X ZP3T2 GMN FROGS.

One male and two female F1 heterozygous GFP+^{eye} ZP3T2γGMN animals were generated as described above. We crossed these F1 animals with our homozygous F1 Ef1αGFPTol2 animals to produce three double transgenic F2 clutches (C1030, C1039, C1040, Figure S1). A subset of each of these clutches harbored both one copy of Ef1αGFPTol2 transposon and ZP3T2γGMN Tol2 transposase constructs (Figure 2). Because the ubiquitous GFP expression of our Ef1αGFPTol2 transposon line masked the GFP eye marker of our ZP3T2γGMN transposase line, any F2 animals that did not produce eye only expression in F3 embryos in an outcross were euthanized. Figure S1 contains detailed information regarding the generation of the three double transgenic lines. Lines are available upon request. If we receive many requests, we will submit the lines to the National *Xenopus* Resource at the Marine Biological Laboratroy in Woods Hole, MA.

RT-PCR analysis of ZP3T2 GMN transposase expression

We isolated total RNA from eggs and stage 30 embryos from outcross matings of GFP+eye F1 ZP3T2γGMN female frogs in each transposase line. We first primed these females with 10 U Human Chorionic Gonadotropin (HCG), and boosted with 200 U HCG approximately 12 hours later. We collected pools of 10 eggs for each sample, then collected pools of 10 stage 30 embryos post-fertilization. We collected testes from GFP+eye F1 adult male frogs in U1946♂ and U1985♀ lines, and used one testis for each sample. We isolated total RNA from all samples using a standard phenol/ chloroform extraction protocol.

We performed first strand cDNA synthesis using Super-Script III First Strand Synthesis System (Invitrogen, Cat. # 18080051) according to manufacturer's protocols. We used 1 μg of RNA per RT reaction. Primers for the ZP3T2γGMN construct (tol2_transposase_F and tol2_transposase_R, Table S1) amplified a 265 bp fragment (Figure 1B). ODC primers (Table S1) were used as a loading control. GFP- embryos, GFP+(-RT) (Figure 1B), and water (data not shown) reactions were used as negative controls.

GFP Expression Analysis

We detected GFP expression using a Zeiss SteREO Lumar.V12 fluorescent microscope and recorded images with a AxioCamHR3 digital camera in conjunction with AxioVision AxioVS V4.8.1.0 imaging software.

Embryos from three F2 double transgenic clutches, C1030, C1039 and C1040, arising from the three transposase injected animals U1946♂, U1984♀ and U1985♀ respectively (Figure S1), were screened for overall number of GFP+ and GFP- embryos, as well as for differences in fluorescent intensity, spatial and temporal GFP expression (Table 1). Examination of the tadpoles did not reveal any changes in spatial or temporal expression, but we did notice that the intensity of the ubiquitous GFP signal varied in a small percentage of tadpoles. We then characterized tadpoles as either High Intensity, Medium Intensity, or Low Intensity, where the majority of tadpoles were of Medium Intensity, similar to the unmanipulated Ef1αGFPTol2 baseline expression. As described in Results (below), we focused our attention on the High Intensity tadpoles. We counted the number of GFP+ tadpoles, identified those that were High Intensity, and reported cumulative totals (Table 1).

Table 1. GFP segregation in tadpoles from double transgenic outcross.

ZP3T2γGMN Line	F2 Tol2/ ZP3T2γGMN	F3 GFP+ to Total Embryos	F3 High GFP+ to Total GFP+ Embryos
U1946 ♂	C1030	1125/2169 (51.9%)*	0/2822 (0%)
U1985 ♀	C1040	404/817 (49.4%)	0/404 (0%)
U1984 ♀	C1039 U2521♂	657/1353 (48.6%)*	11/957 (1.15%)
	C1039 U2522♂	2959/6038 (49.0%)*	78/5576 (1.40%)
	C1039 U2644♀	386/554 (69.7%)	7/386 (1.81%)
	C1039 U2645♀	492/903 (54.5%)	7/492 (1.42%)
	C1039 U2646♀	921/1763 (52.2%)	6/921 (0.65%)
	C1039 U2647♀	928/1713 (54.2%)	19/928 (2.04%)

Column 1: ZP3T2γGMN injected animals, U1946 ♂ U1985 ♀ and U1984 ♀ were used to create three double transgenic F2 lines (C1030, C1040, C1039 respectively) as described in Methods and Figure S1. Column 2: We scored these three F2 double transgenic clutches

for GFP phenotypes. C: Clutch identification number. U: Unique animal identification number. Column 3: Ratio of total ubiquitous GFP+ embryos to total number of embryos from outcross of double transgenic frogs. Column 4: Ratio of High Intensity embryos to total number of GFP+ embryos from outcross of double transgenic frogs. High Intensity embryos were only found in C1039 animals.

For clutches C1030 and C1040, the results in Columns 3 and 4 are the cumulative totals from outcrosses of a number of animals (C1030: 1 ♂ and 8 ♀ C1040: 1 ♂ and 5 ♀). For clutch C1039, results in Columns 3 and 4 are the cumulative totals from multiple outcrosses of each unique animal except U2644 where results are from a single outcross only. In the three cases indicated by * we determined the ratio of GFP+ embryos to total number of embryos from a subset of all the embryos scored. We then identified GFP+ embryos from all embryos collected. From this larger set of GFP+ embryos, we identified High Intensity embryos and calculated the ratio of High Intensity GFP+ embryos to total GFP+ embryos (Column 4). As a result, the denominator in the fourth column is larger than the numerator in the third column.

Transposon Integration Site Analysis

Extension Primer Tag Selection Linker Mediated PCR (EPTS LM-PCR).

We performed integration site analysis on left and right Tol2 transposon arms using an EPTS LM-PCR protocol, as previously described [35]. Briefly, a frequently cutting endonuclease is used to cut genomic DNA close to the transposon arm. A biotinylated primer complementary to the transposon arm is hybridized to the fragments, incubated with septavidin beads, and isolated after multiple washes on a capture magnet. Oligos are ligated onto the genomic end and then the transposon-genomic junction is amplified with primers specific to the transposon arm and ligated oligo. We modified the published protocol by using the restriction enzymes MspI and NlaIII in combination with different primers designed on the Tol2 right arm (Figure 1C, Table S1) to amplify the right arm transposon-genomic junction. LM-PCR products were either sequenced directly (using primers Tol2 5′ N1 and N1-R or Seq1-R, Table S1), or cloned into TOPO-TA vector (Invitrogen, Carlsbad, CA, USA) and then sequenced using standard M13 forward and reverse primers in the vector.

We performed EPTS LM-PCR integration site analysis using DNA from individual stage 40 whole F3 and F4 tadpoles, and using DNA from toe clip tissue of adult F3 High Intensity animals (Figure 2) prepared with the Qiagen DNeasy Blood and Tissue kit "Purification of Total DNA from Animal Tissues" protocol. We did not test any somatic tissue for evidence of somatic remobilization.

Genomic PCR (gPCR).

If we could identify the insertion site of only one transposon arm by LM-PCR, we attempted to amplify the other transposon arm based on its presumed genomic locus ascertained from the successfully cloned arm. We used genomic (Table S1) and transposon (Tol2 5′ N2 and Seq1-R,Table S1) primer combinations to amplify sequence at both left and right transposon arm junctions. We designed genomic primers based on sequence flanking the left and right transposon junction of each putative insertion where possible. In some cases, repetitive sequence made designing primers problematic.

We used the following PCR conditions: denaturation at 94°C for 2 min, followed by cycles of 94°C for 15s, 30s at annealing temp, followed by an elongation step at 72°C. For expected PCR products less than 500 bp, the elongation time was 30s, for 500-1000 bp 60s, and for >1000 bp 90s. Initially we used an annealing temperature of 55°C, and then tried 53°C or 52°C if there was no amplification.

We examined PCR amplification products by gel electrophoresis and then isolated and purified PCR products using the Qiagen Gel Purification kit. We cloned purified PCR products into TOPO-TA before sequencing. Sequences were blasted against the Joint Genome v. 7.1 (http://www.xenbase.org). In some cases, we were unable to clone the integration site due to failure to amplify the integration site or repetitive sequence that made identifying a unique site impossible.

Results

Test for Transposase Expression

We first tested F1 ZP3T2γGMN transgenic frogs from each line for transposase expression by RT-PCR (Figure 1B). In the U1946♂ line, expression was detected in eggs and GFP+eyeembryos from the female frogs and in the testis of adult male frogs. We did not detect expression in eggs from female frogs in the U1984♀ line, but expression was present in GFP+eye embryos. In the U1985♀ line, we did not detect expression in the egg or embryo of female frogs, but found expression in testis of adult male frogs (Figure 1B).

Screen Double Transgenic Offspring for GFP Expression

To ensure the GFP+ and GFP- embryos were segregating at expected 1:1 frequencies, we outcrossed double transgenic animals from each transposase line as described in Methods, and scored for overall number of GFP+ embryos (Table 1, Column 3). Because the ubiquitous GFP expression of our Ef1αGFPTol2 transposon line masked the GFP eye marker of our ZP3T2γGMN transposase line, any F2 animals that did not produce eye only expression in an outcross were euthanized. In cases where the transposase marker was present, overall ubiquitous GFP+ phenotype frequencies reflected simple Mendelian inheritance of the Ef1αGFPTol2 transposon in the F3 generation in most cases. In the case of U2644♀ (from C1039 in the U1984♀ line), the higher than expected percentage of GFP+ embryos (69.7%) is likely due to the presence of bright maternal GFP.

We next scored each outcross for GFP expression differences in GFP+ embryos. We identified three ubiquitous GFP+ fluorescent intensities, categorized as High, Medium, and Low Intensity (Figure 1D). High Intensity embryos were very bright and easily identified visually. Subtle differences in intensity were found among the High Intensity embryos, but were all categorized as High Intensity for the purposes of this study. The intensity of the Medium Intensity embryos was identical to that of their double transgenic parents. Low Intensity embryos showed the least fluorescence. We did not note any change in the GFP+ expression intensity or pattern in the embryos over the observed time period of fertilization to stage 40.

The U1984♀ line is the only line in which we found the High Intensity phenotype. We screened 2 male and 4 female frogs in this line (Table 1, Column 2). All three phenotypes were present in embryos from each mating, with the majority being Medium Intensity, and a small percentage having Low (avg. 2.0%) and High Intensity expression (0.65-2.04%) (Table 1, Column 4).

We screened 1 male and 8 female frogs in the U1946♂ line. The majority of embryos were Medium Intensity, and a small percent were Low Intensity (data not shown). In the U1985♀ line, we saw only Medium Intensity expression in screening 1 male and 5 female double transgenic frogs. We did not find evidence of High Intensity embryos in the U1946♂ or U1985♀ lines (Table 1, Column 4).

Test for Remobilization

Embryos of each intensity (Low, Medium, High) seen in the F3 generation were tested for transposon remobilization by LMPCR +/- gPCR (Figure 2). We found evidence of remobilization only in High Intensity embryos, observed solely in the U1984♀ line (Figure S1). We found apparent remobilization in an average of 1.3% of embryos the U1984♀ line (Table 1, Column 4), and we were able to successfully map new integrations in 67% of these embryos (Table 2). We tested 107 embryos collected from 2 male and 4 female double transgenic frogs from the U1984♀ line. Remobilized transposons were found in the offspring of both male and female frogs. We identified 45 novel PCR products in 42 High Intensity embryos. All successfully mapped High Intensity embryos showed both the original transposon and a new integration. 29 new integrations were mapped on both transposon arms, including confirmation of the TSD (Table 2). One integration was successfully mapped on one arm only; the second arm was not mapped due to repetitive sequence (Table 2, U2646F HI#2). Lack of quality sequence or repetitive sequence on both arms prevented mapping to any loci for 12 samples. In three cases, we were unable to obtain any PCR product. No new PCR products were seen in the 23/23 Low Intensity and 42/42 Medium Intensity embryos tested in the U1984♀ line.

Table 2. Integration site analysis. Row 1: F1 Tol2 (Ef1αGFPTol2) is the original donor transposon insertion site. Novel amplicons from High Intensity F3 offspring from six F2 double transgenic founders (Ef1αGFPTol2/ZP3T2γGMN) are listed in subsequent rows. Column 1: F2 Double transgenic founder animal. Column 2: Novel amplicons from F3 HI expressing offspring. Columns 3 and 4: Left and right flanking genomic sequence at transposon insertion site are in uppercase. Genomic Target Site Duplication (TSD) sequence is in bold. Transposon sequence is in lowercase italics. Column 5: Genomic insertion site (JGI Genome v. 7.1), with chromosome number: base pair position. Column 6: Most proximal flanking gene. Column 7: Distance to most proximal flanking gene (kbp). 1.One arm successfully maps, but sequence flanking other transposon arm is poor quality or not unique.2.Predicted gene.

Parent	F3	Genomic: LA Tol2 sequence	Genomic: RA Tol2 sequence	JGI (v.7.1)	Flanking gene	kbp to gene
F1 Tol2		..ACCTGGAAC**CCCGAAAT**cagaggtgta..	..tacacctctg**CCCGAAAT**CCGCAGACT..	8:87633873	LOC1001251672	intron 1
F2 U2521♂	HI #1	.. TAAGATATA**ATTACCCT**cagaggtgta..	.. tacacctctg**ATTACCCT**TATTGGATG..	2:132651708	Xetro.B02176	126.7
	HI #2	..AGCACGACC**CACACATC**cagaggtgta..	..tacacctctg**CACACATC**ATTGTCAAT..	8:81162239	slc38a6	45.4
	HI #3	..CCTTGCAAA**CATCTGTC**cagaggtgta..	..tacacctctg**CATCTGTC**CATTGGGAA..	1:124132951	LOC4947062	4.4
F2 U2522♂	HI #4	..GTCATTTGC**CTTATACT**cagaggtgta..	..tacacctctg**CTTATACT**GAAGTTTGC..	1:14294794	trappc13	intron 7
	HI #5	..GTGTGCGAC**GTCAGCAC**cagaggtgta..	..tacacctctg**GTCAGCAC**GCACAGGGC..	505:45386	Xetro.K02900	4.1
	HI #7	..TGTCTATT**GTCTTTAGA**cagaggtgta..	..tacacctctg**TCTTTAGA**TCACCTAAT..	8:83640859	lgals3	0.3
	HI #12a	..TTAAAATCC**CTTATTAG**cagaggtgta..	..tacacctctg**CTTATTAG**GGCCACACT..	8:30665922	utp14a	1.3
	HI #13	..AGAGGGTAT**AGGAGTAG**cagaggtgta..	..tacacctctg**AGGAGTAG**GTTAAGGAT..	8:84762850	ubr1	1.8
	HI #14	..TGGGCATTG**GTTTCAGT**cagaggtgta..	..tacacctctg**GTTTCAGT**CATAGGGCT..	8:48893707	gng8	2.9
	HI #16	..CCCTTCTCC**ATTAAGGA**cagaggtgta..	..tacacctctg**ATTAAGGA**GATCCCCCC..	3:47246149	anapc13.2	16.9
	HI #18	..GCCTATTGA**CACACAGC**cagaggtgta..	..tacacctctg**CACACAGC**ACTGTGCAA..	6:9467577	has2	18.7
	HI #19	..TTGGCCTTG**GTGATTAC**cagaggtgta..	..tacacctctg**GTGATTAC**TGATAAGCA..	8 87693428	dpf3	29.3

Parent	F3	Genomic: LA Tol2 sequence	Genomic: RA Tol2 sequence	JGI (v.7.1)	Flanking gene	kbp to gene
	HI #20	..GAAAGATCC**TT-TAAATT**cagag-gtgta..	..tacacctctg**TTTA-AATT**AACTTT-TAG..	8:84932662	stard9	intron 10
	HI #21	..TATTTAGCA**C-GCAAAAG**cagag-gtgta..	..tacacctctg**C-GCAAAAG**T-TATAGTGA..	8:87933146	rgs6	30.5
	HI #22	..CTCGTATCC**AT-TATAAT**cagag-gtgta..	..tacacctctg**AT-TATAAT**ATAT-GCTGT..	8:88268806	pcnx	intron 23
	HI #23	..ATTAATCTA**TT-GATGGC**cagag-gtgta..	..tacacctctg**TT-GATGGC**CTA-ATTCTG..	8:86090368	cfl2	intron 1
	HI #24	..AAATCGATA**CT-CAGAAG**cagag-gtgta..	..tacacctctg**CT-CAGAAG**TCAG-CACTG..	1:86927815	nfil3	5′ UTR intron
	HI #25a	..ACAGTGCAA-**CATATACT**cagag-gtgta..	..tacacctctg-**CATATACT**-TATATTTAA..	184:11129	slc12a4	51.8
	HI #25b	..TTTAGAGA-**TATAATGAT**-cagaggtgta..	..tacacctctg**ATAAT-GAT**ATTTTTCAA..	4:100939936	klf17	24.7
	HI #26	..AAA-CACCTG**GGTAC-CGC**cagaggtgta..	..tacacctctg**GG-TACCGC**AGGTC-TAAA..	9:17814879	sp5	2.2
F2 U2644♀	HI #3	..AGTTAAAC-**CACCTTA-AG**cagaggtgta..	..tacacctctg**ACCT-TAAG**TGC-GACTTG..	8:88132604	sipa1l1	5′ UTR intron
	HI #5	..ACAGTCGC-**CATCTTGCT**-cagaggtgta..	..tacacctctg**ATCTT-GCT**ACAATG-TAA..	8:87633302	Xetro. H01750.1	16.2
	HI #6	..CCTATTTGT**A-ACCCCTG**cagag-gtgta..	..tacacctct-g**AACCCCTG**-GAACATTTT..	8:76909153	klc1	10.8
F2 U2645♀	HI #2	..ACATACACA-**CACACAGC**-cagaggtgta..	..tacacctctg**CACA-CAGC**ATCAT-GTGA..	8:49467657	mark4	intron 1
	HI #3	..GCTACGTG-**TATGGCCAC**-cagaggtgta..	..tacacctctg**ATGGC-C**ACCTTAAATCT..	8:82067778	daam1	5′ UTR intron
	HI #4	..AAGTGTGGT-**CACCTTGG**cagag-gtgta..	..tacacctctg-**CACCTTGG**CTG-CAGTTC..	8:53558200	Xetro.H01082	6.8
	HI #5	..TTTGTAT-CA**ATA-CAACC**cagag-gtgta..	..tacacctctg**ATA-CAACC**ATATA-AAGG..	8c:4355834	Xetro.K05083	9.0
	HI #6b	..AATAGTTGA**G-GAGCAAC**cagag-gtgta..	..tacacctctg**GGAG-CAACA**-CAAGCATG..	7:99952578	dpb	intron 2
F2 U2646♀	HI #21	..GGGG-TATTA**GTG-GCAAA**cagag-gtgta..		8:83208318	tmem260	intron 3
F2 U2647♀	HI #1	..CTGTCATA**CT-GAAACAG**cagag-gtgta..	..tacacctctg**TGAAA-CAG**TGACTA-AGT..	1:165003122	Xetro.A02812	137.2

Remobilizations found relatively close to the donor insertion, or on the donor chromosome, are frequently referred to as "local hops" [23,24,27]. 60% (18) of confirmed remobilizations in our study were on the same scaffold as the parental insertion, and ranged in distance from 571 bp to 83.3 Mb to the donor insertion, with 50% being within 5 Mb of the donor insertion, and 22% within 1 Mb of the donor (Figure 3A). The remaining 40% (12) of remobilizations were found on non-donor scaffolds (Figure 3A).

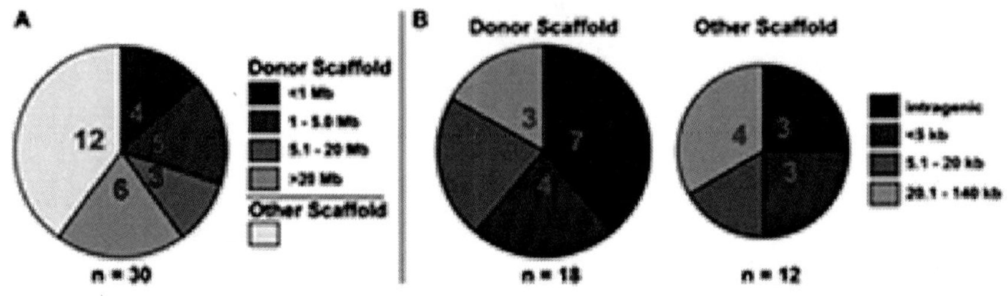

Figure 3. Remobilization Data. A. Number of remobilizations mapped to donor scaffold 8 compared to those mapped to other scaffolds. Proximity (Mb) of remobilized transposons on scaffold 8 to the donor locus is also shown. B. Number of intragenic versus intergenic integrations for remobilizations on both donor and other scaffolds, including proximity (kb) of intergenic integrations to the nearest flanking gene. The size of the pie charts indicates the relative number of remobilizations on donor versus other scaffolds. Number of samples in each category are shown within the pie slices for both A and B, and total n numbers are indicated beneath each graph.

doi:10.1371/journal.pone.0076807.g003

We found that 61% (11) of remobilizations on the donor scaffold were intragenic or less than 5 kb from a gene (Figure 3B, Table 2), a distance that may encompass a gene's regulatory region [24]. The high percentage of these new insertions in close proximity to a gene may be due to the relatively gene rich area in which the donor transposon is located. 50% (6) of remobilizations on non-donor scaffolds were also in or within 5 kb of a gene (Figure 3B, Table 2). 90% (9) of intragenic remobilizations were in introns in the 5′ region of the gene. Though none of the intragenic remobilizations are in exons, two were less than 500 bp away from an exon (Table 2, U2522M HI#22, U2645F HI#2).

We did not find evidence of remobilized transposons in any embryos from the U1946♂ or U1985♀ lines. We tested 14 Low Intensity and 68 Medium Intensity embryos from 1 male and 8 female double transgenic frogs in the U1946♂ line. No evidence of reintegration was found in any of the embryos, and this line was not tested further (Figure S1). In the U1985♀ line, we tested 46 Medium Intensity embryos collected from 1 male and 5 female double transgenic frogs, and found no evidence of remobilization (Figure S1). As these animals also showed little transposase expression (Figure 1B), this line was subsequently culled.

Germline Transmission

To confirm stable germline transmission of the reintegrated transposons, F3 High Intensity embryos from F2 double transgenic male U2522♂ in the U1984♀ transposase line were raised to adulthood (Figure 2). Genomic DNA was collected from the adult F3 frogs, and tested for transposon remobilization by LM-PCR (Table 2, samples U2522M HI# 12a -26). We selected three of these F3 animals (U2625♂, U2634♀, U2635♂, Figure S1), each harboring the donor transposon and one remobilized transposon, to outcross (Table 3, Column 2). Two of these frogs, U2625♂ and U2634♀, had remobilized transposons on non-donor chromosomes (Table 2, U2522M HI#24 and HI#18 respectively, and Table 3, Column 3). The third animal, U2635♂, harbored the donor insertion and a local remobilization on scaffold 8, 635 kb from the donor insertion (Table 2, U2522M HI#22 and Table 3, Column 3).

Table 3. Germline Transmission of Remobilized Transposons in U1984♀ line.

F2 Double Transgenic	F3 High GFP+	F3 Genotype	F4 GFP+ Intensity	F4 GFP+/ total embryos	F4 Sequenced/ Confirmed	F4 Genotype
C1039 U2522 ♂	U2625♂	S8, S1	Low	50/154 (32%)	3/3	S8
			Medium	35/154 (23%)	2/2	S1
			High	41/154 (27%)	3/3	S8 and S1
	U2634♀	S8, S6	Medium	67/168 (40%)	6/6*	S8 or S6
			High	62/168 (37%)	3/3	S8 and S6
	U2635♂	S8, S8□	High	116/241 (48%)	2/2	S8 and S8□

Column 1: We tested germline transmission in F2 double transgenic (Ef1αGFPTol2/ZP3T2γGMN) U2522♂ (see Figure S1), whose F3 embryos showed remobilization (Table 2). We raised F3 GFP+ High Intensity embryos from U2522♂, and confirmed remobilization in these animals by LM-PCR. Columns 2 and 3: We outcrossed three of these F3 animals, U2625♂, U2634♀, and U2635♂, each harboring the original donor transposon, as well a new remobilized insertion. Columns 4 and 5: We scored F4 embryos from each outcross for different GFP+ intensities, as well as ratio of those intensities to overall number of embryos. Column 6: We sequenced the transposon insertion sites of a number of embryos of each intensity to confirm stable germline transmission of the insertions found in the F3 parent. Column 7: In two cases, U2625♂ and U2634♀, transposons segregated independently following simple Mendelian inheritance. In the final case, U2635♂, transposons appeared linked and were in fact located on the identical chromosome. C: Unique clutch identification number. U: Unique animal identification number. * 4/6 embryos of the Medium GFP+ phenotype in the U2634♀cross were confirmed to have the original insertion on S8, and 2/6 were confirmed to have the insertion on scaffold 6. S8 : Original donor insertion on scaffold 8. S8□ : Remobilized insertion on donor scaffold 8.

We scored F4 embryos from outcrosses of these F3 High Intensity animals for fluorescent expression level and for numbers of embryos of each expression level present (Table 3, Columns 4 and 5). We found embryos from U2625♂ showed three expression levels, High, Medium and Low Intensity, while embryos from U2634♀ had two expression levels, Medium and High Intensity. Embryos from both U2625♂ and U2634♀ had different intensities present in percentages consistent with two independently assorting transposon insertions, as expected since the remobilized transposon was on a non-donor chromosome (Table 3, Columns 4 and 5). We observed only High Intensity F4 embryos from U2635♂, whose remobilized insertion was found to be 635 kb from the original insertion on scaffold 8. In the case of transposon insertions linked by close proximity, all of the offspring would inherit both insertions, and would be expected to have the same phenotype (Table 3, Columns 4 and 5).

We performed EPTS LM-PCR on a number of embryos of each expression level in the F3 outcrosses described above (Table 3, Columns 4, 6 and 7). This confirmed the correlation of the Low and

Medium Intensity expression to embryos with single insertions, and the High Intensity expression to embryos inheriting both insertions. Of note, in the case of U2634♀, embryos inheriting only one transposon insertion (the original insertion on scaffold 8 or the remobilized transposon on scaffold 6), each showed an identical Medium Intensity fluorescence. We tested 6 of these embryos by LM-PCR, and found that 4 embryos harbored the original insertion, and 2 had inherited the new insertion on scaffold 6.

Discussion

In this study, we show a substrate Tol2 transposon can be remobilized by crossing to transposase expressing animals. We believe this strategy is a viable and efficient method of screening for new insertions when those embryos showing increased fluorescence (High Intensity) are selected for testing. This phenotype was easy to identify in a dish of embryos, and we were able to successfully map reintegrations in the majority of these embryos. Given the remarkable fecundity of X. tropicalis, even a relatively low remobilization rate can result in a substantial number of remobilized transposons. All of the successfully mapped High Intensity F3 embryos showed both the original transposon and the new integration. A similar result has been found in other studies of Tol2 remobilization, and has been postulated to be a result of transposon duplication in the S-phase of mitosis [22,24].

While we did not find any evidence of reintegration in Low and Medium Intensity F3 embryos, it may be occurring rarely enough to be difficult to detect without screening very large numbers of embryos. Thus, it is possible remobilization rates are higher than we have found, but random selection of embryos has proven to be an inefficient method of finding reintegrations. Overall, our apparent remobilization rate using the breeding based method was relatively low, a range of 0.65-2.04% among different F2 animals (Table 2), an average of 1.3%. A breeding based remobilization strategy using the Sleeping Beauty transposon system [27] in Xenopus showed similar efficiency to our study in producing reintegration in offspring, less than 1% on average. An injection based Tol2 remobilization strategy [22] resulted in a very high remobilization efficiency, where 18/18 injected embryos analyzed showed evidence of at least one new integration. These results are similar to those found in Tol2 remobilization studies in zebrafish, where remobilization by injection resulted in relatively high apparent germline remobilization rates (8.3-100% among different animals in one study [23] and 38% and 48% from two screens in another [24]). In contrast, in vivo supplied transposase resulted in a much lower rates of remobilization in zebrafish. Transposase supplied in vivo using a heat shock promoter was not able to induce remobilization at early embryonic stages, but at adult stages after multiple heat shocks, Tol2 was able to remobilize at an overall rate of 6.2% [23].

Thus while the injection based method is relatively labor intensive, as transposase mRNA is prepared and embryo injections performed each time remobilization is desired, the observed remobilization rate is presently higher. The breeding based strategy, while currently less efficient at inducing remobilization, is less labor intensive, only requiring setting up a mating, then collecting and analyzing resulting embryos. Thus, it would be desirable to increase the efficiency of the breeding based system to combine the benefits of both higher remobilization efficiency and reduced labor.

A number of variables have potential to be optimized to increase efficiency in the Tol2 breeding based remobilization system. Increasing the number of Tol2 donor insertions would provide more substrate transposons for remobilization. In the Sleeping Beauty breeding based study [27], a higher overall apparent remobilization rate was seen in animals from a line bearing a concatamer of 8-10 transposons (~ 1%) than in animals from a line with a concatomer of only 3 transposons (~ 0.2%). The higher rate of remobilization was partially attributed to the greater number of substrate transposons. We chose to use animals bearing one Tol2 transposon insertion in order to simplify analysis of remobilization in this study. Animals harboring multiple canonical insertions can readily be created by injection or breeding based methods to potentially increase remobilization efficiency in future studies.

More precise titration of the transposase dose necessary to induce remobilization could increase remobilization efficiency. Southern blots to determine transposase transgene copy number or quantitative PCR to determine transposase levels more precisely in our remobilized line would be beneficial in this regard. In addition, efficiency could be improved with a "next generation" transposase, similar to the "hyperactive" Sleeping Beauty transposase SB100X, which has 100 times the activity of the original SB10 form [36].

Choice of promoter is another variable that could affect efficiency of remobilization. A sperm specific promoter would be ideal for maintaining stable remobilized transgenic lines, as maternal transposase protein produced by female frogs harboring transposase contructs with an egg or ubiquitous promoter would likely cause continued remobilization in subsequent generations, preventing the stable transmission of remobilized transposons.

When developing our approach, we attempted to create a sperm specific transposase line, but this became technically challenging and unsuccessful. We created transposase animals containing somatic (Ef1α) expression constructs, but these have yet to be tested. We chose to focus on the zp3 promoter, as we were successful in generating the lines, and we reasoned that some tissue specificity might be beneficial. zp3 has been reported as an egg specific glycoprotein in a number of species including zebrafish, human and mouse [31]. We thus expected egg specific transposase expression from our construct. Our RT-PCR results indicated expression not only in the egg, but in embryos of female frogs and testis from adult male frogs (Figure 1B). –RT controls indicate the result is not due to contamination of the samples. These results suggest perhaps the zebrafish zp3 promoter is functioning less specifically in frog.

In addition to finding non-oocyte transposase expression, we also found relative differences in the amounts of transposase expressed among the three ZP3T2γGMN lines (Figure 1B). Different transposase transgene insertion loci, insertion copy numbers in the concatemer, and a variety of local host factors may result in differing expression levels among the different transposase lines, thus affecting the efficiency of transposon remobilization. Interestingly, in spite of the relatively high transposase expression observed, the U1946♂ line produced no detectable remobilizations. It is possible that this line was affected by "overproduction inhibition", in which transposase excess over a threshold level has been shown to inhibit the excision of transposons. This is known to occur most notably in Sleeping Beauty, but has also recently been shown for Tol2 transposase in human Hela cells [25,26]. Since our results suggest that expression levels are important for Tol2 mobilization, quantitative PCR may be useful for further optimizing remobilization. However, empiric testing is likely to remain necessary since the relationship between mobilization and expression levels is complex. Southern blot analysis to determine the number of insertions present in the concatemers of the three lines may be useful in guiding development of efficient transposase lines for future studies.

Remobilizations found on the donor chromosome, or relatively close to the donor insertion, are frequently referred to as "local hops" [23,24,27], and are desirable when attempting to saturate a genic region in gene or enhancer trapping strategies. Previous studies in frog and other species have reported local hopping of the Tol2 transposon, with multiple remobilizations less than 5 Mb from the donor insertion [22-24,37]. The injection based study of Tol2 remobilization in X. tropicalis reported 20% of remobilized transposons integrated near the donor locus [37]. In zebrafish, an average 14-20% local hopping frequency was found in a Tol2 remobilization study in which different donor loci were used, as well as different methods of supplying transposase (mRNA injection, single or multiple heat shocks) [23]. In contrast, 80% of Sleeping Beauty transposons were found to integrate within 3 Mb of the original insertion [27], consistent with previous reports that Sleeping Beauty tends to remobilize locally [38]. In our study, we found 60% (18) of confirmed remobilizations were on the same scaffold as the parental insertion, with 50% being within 5 Mb of the donor insertion, and 22% within 1 Mb of the donor (Figure 3A). This frequency is higher than previous rates reported for Tol2 and closer to rates reported for Sleeping Beauty.

Tol2 has also been reported to favor insertion into transcriptional units and 5′ genic areas [24-26], a highly desirable feature for enhancer and gene trapping strategies. This is in contrast to Sleeping Beauty, which has been reported to favor intergenic regions and shows little tendency to target 5′ gene regions [26]. One study of Tol2 reintegration in zebrafish has suggested an arbitrary 5 kb interval at the 3′ and 5′ ends of genes as encompassing a gene's regulatory region [24]. We found a high percentage, 61% (11), of local remobilizations were within or less than 5 kb from a gene (Figure 3B, Table 2). While this may be partially attributed to the relatively gene rich area in which the donor transposon is located, we also found 50% (6) of remobilizations on non-donor scaffolds were in or within 5 kb of a gene (Figure 3B, Table 2). Additionally, 90% (9) of remobilizations actually within a gene were in introns in the 5′ region of the gene (Figure 3B, Table 2). Thus, our findings again suggest that breeding based strategies using Tol2 will have applications in enhancer and gene trapping studies.

Future breeding based remobilization strategies could potentially take advantage of the differences in transposon systems such as Tol2 and Sleeping Beauty with complementary or synergistic approaches to enhancer or gene trapping. For instance, a transposon construct in which a Sleeping Beauty cassette bearing the desired cargo is nested in a Tol2 construct could be used to insert and remobilize insertions to genic areas via Tol2 transposase, then the high tendency for local hopping of Sleeping Beauty could be exploited to locally saturate the area for increased enhancer or gene trapping.

SUMMARY

Our results demonstrate Tol2 transposon remobilization can occur in *X. tropicalis* using a breeding based strategy, and the re-integrated transposons are stably transmitted through the germline. This method could be used for enhancer trapping to create useful reporter lines. Improvements to the method, such as an increase in the number of substrate transposons, more precise titering of transposase expression, or a more efficient transposase could improve the remobilization efficiency. In the future, by selecting a transposon with relatively weak basal expression (a minimal promoter), remobilization could trap local enhancers, generating localized, specific GFP expression that may prove particularly useful for different experimental applications.

We found remobilized Tol2 transposons frequently landed near or in genes, a factor which could facilitate the trapping of genes or enhancers. Additionally, remobilizations showed a tendency to land close to the donor insertion, which further benefits gene trapping when the donor transposon is in a gene rich area, or near a gene of interest. With refinements and further work, this breeding based strategy could generate many useful transgenic lines in *X. tropicalis*, further advancing this already useful model.

ACKNOWLEDGMENTS

The authors thank Sarah Kubek, who assisted in mating and screening animals, and Michael Slocum, who performed husbandry to raise and optimize growth of animals. Khokha Lab members, especially Florencia del Viso, kindly provided comments and input for experiments and manuscript.

Author Contributions

Conceived and designed the experiments: MAL MK MKK. Performed the experiments: MAL MK MKK. Analyzed the data: MAL MKK. Contributed reagents/materials/analysis tools: MAL MK MKK. Wrote the manuscript: MAL MK MKK.

REFERENCES

1. Hirsch N, Zimmerman LB, Grainger RM (2002) Xenopus, the next generation: X. tropicalis genetics and genomics. Dev Dyn 225: 422-433. doi:10.1002/dvdy.10178. PubMed: 12454920.
2. Wallingford JB, Liu KJ, Zheng Y (2010) Xenopus. Curr Biol 20: R263-R264. doi:10.1016/j.cub.2010.01.012. PubMed: 20334828.
3. Offield MF, Hirsch N, Grainger RM (2000) The development of Xenopus tropicalis transgenic lines and their use in studying lens developmental timing in living embryos. Development 127: 1789-1797. PubMed: 10751168.
4. Yergeau DA, Mead PE (2007) Manipulating the Xenopus genome with transposable elements. Genome Biol 8 Suppl 1: S11. doi:10.1186/gb-2007-8-s1-s11. PubMed: 18047688.
5. Harland RM, Grainger RM (2011) Xenopus research: metamorphosed by genetics and genomics. Trends Genet 27: 507-515. doi:10.1016/j.tig.2011.08.003. PubMed: 21963197.
6. Hirsch N, Zimmerman LB, Gray J, Chae J, Curran KL et al. (2002) Xenopus tropicalis transgenic lines and their use in the study of embryonic induction. Dev Dyn 225: 522-535. doi:10.1002/dvdy.10188. PubMed: 12454928.
7. Clark KJ, Geurts AM, Bell JB, Hackett PB (2004) Transposon vectors for gene-trap insertional mutagenesis in vertebrates. Genesis 39: 225-233. doi:10.1002/gene.20049. PubMed: 15286994.

8. Korzh V (2007) Transposons as tools for enhancer trap screens in vertebrates. Genome Biol 8 Suppl 1: S8. doi:10.1186/gb-2007-8-1-r8. PubMed: 18047700.
9. Doherty JR, Johnson Hamlet MR, Kuliyev E, Mead PE (2007) A flk-1 promoter/enhancer reporter transgenic Xenopus laevis generated using the Sleeping Beauty transposon system: an in vivo model for vascular studies. Dev Dyn 236: 2808-2817. doi:10.1002/dvdy.21321. PubMed: 17879322.
10. Balciunas D, Ekker SC (2005) Trapping fish genes with transposons. Zebrafish 1: 335-341. doi:10.1089/zeb.2005.1.335. PubMed: 18248211.
11. Clark KJ, Balciunas D, Pogoda HM, Ding Y, Westcot SE et al. (2011) In vivo protein trapping produces a functional expression codex of the vertebrate proteome. Nat Methods 8: 506-515. doi:10.1038/nmeth.1606. PubMed: 21552255.
12. Kawakami K (2005) Transposon tools and methods in zebrafish. Dev Dyn 234: 244-254. doi:10.1002/dvdy.20516. PubMed: 16110506.
13. O'Kane CJ, Gehring WJ (1987) Detection in situ of genomic regulatory elements in Drosophila. Proc Natl Acad Sci U S A 84: 9123-9127. doi:10.1073/pnas.84.24.9123. PubMed: 2827169.
14. Liao HK, Wang Y, Noack Watt KE, Wen Q, Breitbach J et al. (2012) Tol2 gene trap integrations in the zebrafish amyloid precursor protein genes appa and aplp2 reveal accumulation of secreted APP at the embryonic veins. Dev Dyn 241: 415-425. doi:10.1002/dvdy.23725. PubMed: 22275008.
15. Calvi BR, Hong TJ, Findley SD, Gelbart WM (1991) Evidence for a common evolutionary origin of inverted repeat transposons in Drosophila and plants: hobo, Activator, and Tam3. Cell 66: 465-471. doi:10.1016/0092-8674(81)90010-6. PubMed: 1651170.
16. Koga A, Suzuki M, Inagaki H, Bessho Y, Hori H (1996) Transposable element in fish. Nature 383: 30. doi:10.1038/383030b0. PubMed: 8779712.
17. Kawakami K, Shima A, Kawakami N (2000) Identification of a functional transposase of the Tol2 element, an Ac-like element from the Japanese medaka fish, and its transposition in the zebrafish germ lineage. Proc Natl Acad Sci U S A 97: 11403-11408. doi: 10.1073/pnas.97.21.11403
18. Koga A (2004) Transposition mechanisms and biotechnology applications of the medaka fish Tol2 transposable element. Adv Biophys 38: 161-180. doi:10.1016/S0065-227X(04)80151-5. PubMed: 15493333.

19. Balciunas D, Wangensteen KJ, Wilber A, Bell J, Geurts A et al. (2006) Harnessing a high cargo-capacity transposon for genetic applications in vertebrates. PLOS Genet 2: e169. doi:10.1371/journal.pgen.0020169. PubMed: 17096595.

20. Hamlet MR, Yergeau DA, Kuliyev E, Takeda M, Taira M et al. (2006) Tol2 transposon-mediated transgenesis in Xenopus tropicalis. Genesis 44: 438-445. doi:10.1002/dvg.20234. PubMed: 16906529.

21. Yergeau DA, Kelley CM, Zhu H, Kuliyev E, Mead PE (2010) Transposon transgenesis in Xenopus. Methods 51: 92-100. doi:10.1016/j.ymeth.2010.03.001. PubMed: 20211730.

22. Yergeau DA, Kelley CM, Kuliyev E, Zhu H, Sater AK et al. (2010) Remobilization of Tol2 transposons in Xenopus tropicalis. BMC Dev Biol 10: 11. doi:10.1186/1471-213X-10-11. PubMed: 20096115.

23. Urasaki A, Asakawa K, Kawakami K (2008) Efficient transposition of the Tol2 transposable element from a single-copy donor in zebrafish. Proc Natl Acad Sci U S A 105: 19827-19832. doi:10.1073/pnas.0810380105. PubMed: 19060204.

24. Kondrychyn I, Garcia-Lecea M, Emelyanov A, Parinov S, Korzh V (2009) Genome-wide analysis of Tol2 transposon reintegration in zebrafish. BMC Genomics 10: 418. doi:10.1186/1471-2164-10-418. PubMed: 19737393.

25. Grabundzija I, Irgang M, Mátés L, Belay E, Matrai J et al. (2010) Comparative analysis of transposable element vector systems in human cells. Mol Ther 18: 1200-1209. doi:10.1038/mt.2010.47. PubMed: 20372108.

26. Muñoz-López M, García-Pérez JL (2010) DNA transposons: nature and applications in genomics. Curr Genomics 11: 115-128. doi:10.2174/138920210790886871. PubMed: 20885819.

27. Yergeau DA, Kelley CM, Kuliyev E, Zhu H, Johnson Hamlet MR, et al. (2011) Remobilization of Sleeping Beauty transposons in the germline of Xenopus tropicalis. Mob DNA 2: 15. doi:10.1186/1759-8753-2-15. PubMed: 22115366.

28. Bellen HJ, O'Kane CJ, Wilson C, Grossniklaus U, Pearson RK et al. (1989) P-element-mediated enhancer detection: a versatile method to study development in Drosophila. Genes Dev 3: 1288-1300. doi:10.1101/gad.3.9.1288. PubMed: 2558050.

29. Bonin CP, Mann RS (2004) A piggyBac transposon gene trap for the analysis of gene expression and function in Drosophila. Genetics 167: 1801-1811. doi:10.1534/genetics.104.027557. PubMed: 15342518.

30. Quiñones-Coello AT, Petrella LN, Ayers K, Melillo A, Mazzalupo S

et al. (2007) Exploring strategies for protein trapping in Drosophila. Genetics 175: 1089-1104. PubMed: 17179094.

31. Liu X, Wang H, Gong Z (2006) Tandem-repeated Zebrafish zp3 genes possess oocyte-specific promoters and are insensitive to estrogen induction. Biol Reprod 74: 1016-1025. doi:10.1095/biolreprod.105.049403. PubMed: 16481590.

32. Kawakami K, Shima A (1999) Identification of the Tol2 transposase of the medaka fish Oryzias latipes that catalyzes excision of a nonautonomous Tol2 element in zebrafish Danio rerio. Gene 240: 239-244. doi:10.1016/S0378-1119(99)00444-8. PubMed: 10564832.

33. Ogino H, McConnell WB, Grainger RM (2006) Highly efficient transgenesis in Xenopus tropicalis using I-SceI meganuclease. Mech Dev 123: 103-113. doi:10.1016/j.mod.2005.11.006. PubMed: 16413175.

34. Nieuwkoop PD, Faber J (1994) Normal table of Xenopus laevis (Daudin) : a systematical and chronological survey of the development from the fertilized egg till the end of metamorphosis. New York: Garland Publishing 210 p., leaves of plates p.

35. Yergeau DA, Kuliyev E, Mead PE (2007) Injection-mediated transposon transgenesis in Xenopus tropicalis and the identification of integration sites by modified extension primer tag selection (EPTS) linker-mediated PCR. Nat Protoc 2: 2975-2986. doi:10.1038/nprot.2007.428. PubMed: 18007633.

36. Mátés L, Chuah MK, Belay E, Jerchow B, Manoj N et al. (2009) Molecular evolution of a novel hyperactive Sleeping Beauty transposase enables robust stable gene transfer in vertebrates. Nat Genet 41: 753-761. doi:10.1038/ng.343. PubMed: 19412179.

37. Yergeau DA, Kelley CM, Zhu H, Kuliyev E, Mead PE (2012) Forward genetic screens in Xenopus using transposon-mediated insertional mutagenesis. Methods Mol Biol 917: 111-127. doi:10.1007/978-1-61779-992-1_6. PubMed: 22956084.

38. Miskey C, Izsvák Z, Kawakami K, Ivics Z (2005) DNA transposons in vertebrate functional genomics. Cell Mol Life Sci 62: 629-641. doi:10.1007/s00018-004-4232-7. PubMed: 15770416.

Chapter 3

LIMITED FITNESS ADVANTAGES OF CROP-WEED HYBRID PROGENY CONTAINING INSECT-RESISTANT TRANSGENES (BT/CPTI) IN TRANSGENIC RICE FIELD

[1]Xiao Yang, Feng Wang, Jun Su, [1]Bao-Rong Lu

[1]Ministry of Education Key Laboratory for Biodiversity Science and Ecological Engineering, Department of Ecology and Evolutionary Biology, Fudan University, Shanghai, China

[2]Fujian Province Key Laboratory of Genetic Engineering for Agriculture, Fujian Academy of Agricultural Sciences, Fuzhou, China

ABSTRACT

Background

The spread of insect-resistance transgenes from genetically engineered (GE) rice to its coexisting weedy rice (*O. sativa f. spontanea*) populations *via* gene flow creates a major concern for commercial GE rice cultivation. Transgene flow to weedy rice seems unavoidable. Therefore, characterization of potential fitness effect brought by the transgenes is essential to assess environmental consequences caused by crop-weed transgene flow.

Methodology/Principal Findings

Field performance of fitness-related traits was assessed in advanced hybrid progeny of F_4generation derived from a cross between an insect-resistant transgenic (*Bt/CpTI*) rice line and a weedy strain. The performance of transgene-positive hybrid progeny was compared with the transgene-negative progeny and weedy parent in pure and mixed planting of transgenic and nontransgenic plants under environmental conditions with natural *vs.* low insect pressure. Results showed that under natural insect pressure the insect-resistant transgenes could effectively suppress target insects and bring significantly increased fitness to transgenic plants in pure planting, compared with nontransgenic plants (including weedy parent). In contrast, no significant differences in fitness were detected under low insect pressure. However, such increase in fitness was not detected in the mixed planting of transgenic and nontransgenic plants due to significantly reduced insect pressure.

Conclusions/Significance

Insect-resistance transgenes may have limited fitness advantages to hybrid progeny resulted from crop-weed transgene flow owning to the significantly reduced ambient target insect pressure when an insect-resistant GE crop is grown. Given that the extensive cultivation of an insect-resistant GE crop will ultimately reduce the target insect pressure, the rapid spread of insect-resistance transgenes in weedy populations in commercial GE crop fields may be not likely to happen.

INTRODUCTION

The commercial application of genetically engineered (GE) crops in agricultural production has aroused great biosafety concerns worldwide. The potential environmental impacts caused by the cultivation of the GE crops are the most debated issues [1]-[3].

Transgene flow from a GE crop into populations of wild or weedy relatives and its potential ecological risks is considered as a key environmental problem [4]–[6]. (Trans) gene flow from a crop to its wild relatives has been widely documented in the last decades [7]–[9]. However, our knowledge on the role of introgressed transgenes that confer novel traits with a strong selective advantage in changing the evolutionary process of wild or weedy populations is still limited. Therefore, assessing potential environmental risks caused by the extensive cultivation of GE crops prior to their commercialization becomes a common practice [1], [5], [8]. The study of potential ecological consequences created by transgene flow to wild relatives particularly the coexisting and conspecific weeds will provide solid bases for environmental risk assessment [10]–[13]. The fate of weedy populations that acquired transgenes through gene flow is largely different, depending on the fitness effect of the introgressed transgenes under given environmental conditions [5],[14]. If the introgressed transgenes can increase fitness, the transgenes will enhance the competitiveness and invasiveness of the weedy populations, leading to the rapid spread of the transgenes in the weedy populations, and *vice versa*. Thus, estimating fitness effect of transgenes on weedy populations is essential.

Rice is an important world crop providing staple food for nearly one half of the global population[15]. Research and development of GE rice has been extensively practiced in China, and consequently a many GE rice lines with novel traits such as insect-resistance, herbicide-tolerance, and improved grain quality have been developed [5], [11], [16]. In 2009, Chinese authority granted biosafety certificates for two insect-resistant rice lines containing a *Bt*transgene (*Bacillus thuringiensis*) [17], meaning that GE rice may enter commercial production in near future. Transgene spread from insect-resistant GE rice to its coexisting weedy rice populations through gene flow becomes a major environmental biosafety concern. Weedy rice (also known as red rice, *O. sativa f. spontanea*) is a noxious weed that causes significant losses of rice yield and quality worldwide [18], [19]. The introgression of transgenes with selective advantages may largely enhance the spread of the weedy rice, causing more serious weed problems. Weedy rice is an annual weed conspecific to cultivated rice [20], and transgene flow from cultivated rice to weedy rice cannot be avoided owing to their similar flowering phenology[21]. It is therefore valid to determine the fitness effect of insect-resistance transgenes on weedy rice populations under different insect pressure for risk assessment.

Fitness of an insect-resistance transgene is largely associated with the environment in which insect pressure can largely be variable [22]–[25]. The relationship between fitness effect brought by an insect-resistance transgene and the ambient insect pressure has not been well described. A number of studies of crop-weed hybrids showed variable fitness effect brought by insect-resistant transgenes under different insect occurrences [25], [26]. It is apparent based on these studies that the enhanced fitness benefit of insect-resistant transgenes was always associated with the high target herbivore pressure, whereas under low herbivore pressure the benefit brought by the insect-resistant transgenes was reduced. Therefore, to analyzed fitness effect brought by an insect-resistance transgene under designed experimental conditions with different insect pressure should be the key to detect the fitness effect of such a transgene.

Studies have already revealed the effects of *Bt* transgene in crop-wild or crop-weed hybrid progeny under different environments [23], [25], [26]–[28]. Our previous studies of F_1–F_3 crop-weed hybrids and their derived lineages from insect-resistant rice revealed increased fecundity brought by *Bt/CpTI* transgenes when target insects were abundant [25], [29]. However, the relationship between variation in different insect pressure and the corresponding fitness change was not well addressed by a properly designed experiment. In this study, we intended to address the following questions: (i) Are the insect-resistant transgenes (*Bt/CpTI*) still effective on resisting rice target herbivores in the transgenic F_4 populations derived from a GE crop-weed hybrid under natural insect pressure? (ii) Does the transgenic F_4 population have significantly higher fitness than the nontransgenic F_4 population and weedy rice parent under natural insect pressure? (iii) Does the mixed planting of transgenic plants significantly reduce the ambient insect pressure of the nontransgenic plants, and consequently, reduce the potential fitness benefit brought by insect-resistance transgenes in the experimental fields? The answer of above questions is essential for assessing the long-term ecological impacts caused by insect-resistant transgene flow from to its weedy rice populations.

METHODS

Experimental Materials and their Sources

Our experiment included two sets of F_4 populations, with or without transgenes derived from a crop-weed hybrid, and its weedy rice parent. The F_4 populations were generated by continued self-pollination from the F_1 hybrids of an insect-resistant GE rice line crossed with a weedy rice strain (used as the pollen recipient) [25], [29]. The GE rice line produced by Fujian Academy of Agricultural Sciences in Fuzhou of China had the *Bt/CpTI* (*Bacillus thuringiensis* (*cryIAc*) and cowpea trypsin inhibitor) transgenes in a double insertion, tightly linked with the selectable marker gene *hyg* (hygromycin resistance), transformed by an agrobacterium method [30]. This GE rice line was produced to control lepidopteran pests such as rice stem borers (*Scirpophaga incertulas, Chilo suppressalis,* and *Sesamia inferens*) and rice leaf-folders (*Cnaphalocrocis medinalis*) [24]. The weedy rice strain (coded as W) was donated by Dr. H. S. Suh of Yeungnam University from S Korea [29]. The identification and separation of the transgene homozygous population (coded as TP) from non-transgene homozygous population (TN) were made from screening the seedlings derived from seeds of F_2 plants using the hygromycin-B water solution treatment [25]. The selected transgene-positive and transgene-negative F_3 populations were self-pollinated for one more generation. Therefore, plants of both transgene-positive and transgene-negative F_4 populations were homozygous at the transgenic locus.

Experimental Design

The field experiment was conducted in 2010 in the environment with a natural insect occurrence at the designated Biosafety Assessment Center in Fuzhou, Fujian Province, China. The experimental plots were allocated in two separate blocks (natural insect *vs.* low insect pressure) isolated from each other for >100 m. The block with natural insect pressure was free of any insecticide treatment, whereas the block with low insect pressure was achieved by spraying different insecticides (Methamidophos, Folimat, Buprofezin, and Monosultap) every 7–10 days starting at the tillering stage. The main comparison was made among the weedy parent, transgenic and nontransgenic hybrid lineages under the different insect pressure. Therefore, under each of the insect-pressure environmental conditions (natural *vs.* low), the following treatments were designated: (1) pure planting of weedy rice parent, transgene-positive, and transgene-negative F_4 populations; (2) mixed planting between transgene-positive and -negative F_4 populations, or transgene-positive F_4 population and weedy rice parent, or transgenic-negative F_4 population and weedy rice parent. Consequently, there were a total of twelve treatments included in the experiment.

For each treatment, four replicates (plots) were included. In the pure planting design, each plot included 64 plants in an 8×8 grid with 20 cm spacing between plants. In the mixed planting design, each plot included 32 plants of either type of the experimental materials grown alternatingly in an 8×8 grid with 15 cm spacing between plants. The field layout of plots was arranged in a complete randomized design in the two blocks. Seeds of all the plant materials were germinated in 50×100 cm pots filled with paddy soil from fields and placed in a green house. Seedlings were transferred to a nursery bed 20 days after seed germination, and then transplanted into field plots ~40 days after seed germination. To avoid accidental seed shattering, panicles of all plants were enclosed by nylon mesh-bags 10–15 days after the flowering.

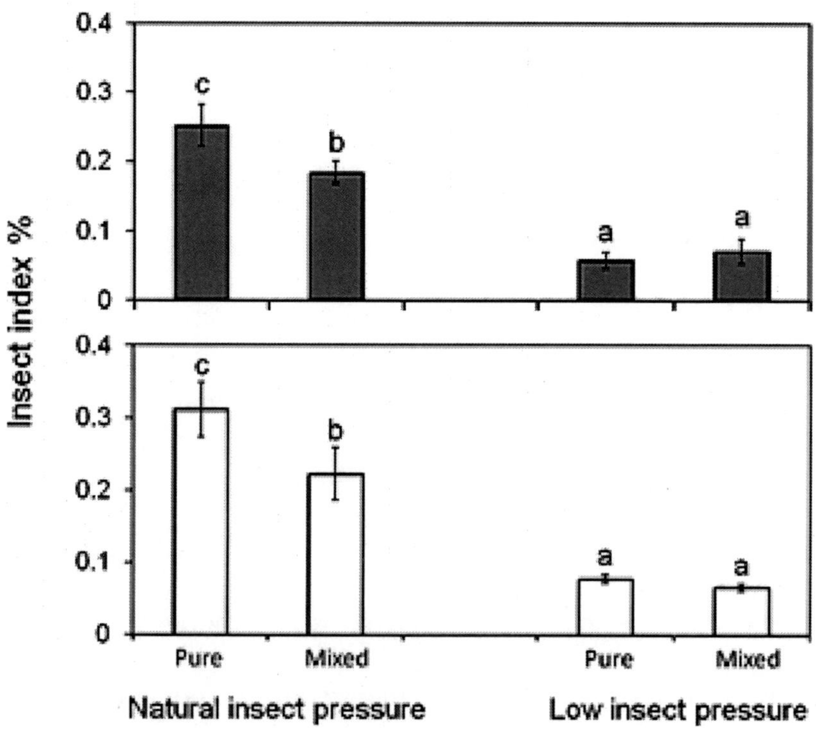

Figure 1. Insect index (%) calculated based on the ratio of blasted tillers and folded leaves on weedy parent (above) and transgene-negative hybrid progeny (below) in different cultivation modes (pure vs. mixed) under natural or low insect pressure.

The comparison was made among pure and mixed planting under natural insect and low insect pressure. Different letters above the columns indicate significant differences according to Duncan's multiple range tests after Bonferroni correction. "Pure" indicates pure planting, "Mixed" indicates mixed planting. Bars represent standard error. Levels of significance: $P<0.05$.

doi:10.1371/journal.pone.0041220.g001

Data Collection and Analysis

For estimating insect damage and fitness-related traits, all plants in a plot were characterized at various vegetative and reproductive stages, except for those in the border rows that were excluded to avoid the edge effect. As a result, a total 36 plants from a pure-planting plot and 18 plants for each of the two types of populations from a mixed-planting plot were characterized. Blasted tillers (by rice stem borers) and folded leaves (by leaf-folder) at the beginning of flowering stage were measured to estimate the insect damage. An insect index was used to estimate the insect pressure that was calculated as the sum of percentage of both blasted tiller and folded leaves in the plots of weedy parent and nontransgenic F_4 populations. The following fitness-related traits were included for measurement: plant height, number of tillers, number of panicles, and number of filled seeds per plant, seed set, and 1000-seed weight at various growth stages.

One way ANOVA (Duncan's multiple range test) was used to examine significant differences for insect index obtained from weedy rice and transgene-negative plants in pure and mixed planting under natural and low insect pressure, and to test significant differences in insect damage and fitness-related traits among weedy parent, transgene-positive, and transgene-negative populations in pure planting, followed by the Bonferroni correction. The paired *t*-test was used to examine significant differences between transgene-positive and transgene-negative, transgene-positive and weedy parental, or transgene-negative and weedy parental populations in mixed planting. All statistical analyses were performed using the software package IBM SPSS Statistics ver. 19.0 for Windows (SPSS Inc., IBM Company Chicago, IL, USA, 2010).

RESULTS

Insect Pressure Under the Two Environmental Conditions

The insect index calculated from the insect damage of blasted tillers and folded-leaves demonstrated a significant difference between the two blocks with natural insect pressure and low insect pressure (Fig. 1; Table 1 and 2). As estimated from the weedy parental and transgene-negative F_4 populations in pure planting, the insect index was significantly higher under the natural insect pressure (25.3%

and 31.5%) than that the under low insect pressure (5.8% and 7.8%; $P<0.001$). The significant differences of insect pressure allowed us to determine the net benefit caused by the transgenes when all the plants were exposed to herbivores, and the net cost caused by the transgenes when herbivores were absent or reduced to a low level.

Table 1. Average values and standard errors (±) of insect damage and fitness-related traits in weedy rice parent, transgene-positive, and transgene-negative crop-weed hybrid progeny (F_4) in pure planting under natural (above) vs. low (below) insect pressure.

doi:10.1371/journal.pone.0041220.t001

Trait	Plant material		
	Weedy parent	Transgene-positive	Transgene-negative
Natural insect			
Blasted tillers %	11.4±1.1 b	7.0±1.3 a	15.6±2.0 b
Folded leaves %	11.8±2.1b	1.5±0.3 a	15.6±3.1 b
Plant height (cm)	91.9±4.0 a	111.9±4.2 b	112.3±2.6 b
No. of tillers	17.3±1.1 a	18.2±0.3 a	16.0±1.2 a
No. of panicles	10.4±1.0 a	15.0±0.5 b	12.0±0.9 a
Seed set (%)	50.0±1.5 b	47.8±2.5 b	40.8±0.9 a
1000-seed weight (g)	20.3±0.2 a	21.4±0.3 b	20.7±0.2 ab
Low insect			
Blasted tillers %	4.8±1.2 a	5.0±0.8 a	4.5±0.7 a
Folded leaves %	1.0±0.0 b	0.2±0.2 a	1.9±0.2 b
Plant height (cm)	107.3±1.7 a	111.0±2.5 a	120.1±2.7 b
No. of tillers	11.9±0.6 a	15.3±0.4 a	15.0±0.9 a
No. of panicles	11.6±0.7 b	10.6±0.7 a	11.0±0.4 a
Seed set (%)	51.3±1.3 ab	52.2±1.4 b	46.0±0.8 a
1000-seed weight (g)	19.7±0.2 a	22.1±0.2 b	21.6±0.3 b

Different letters following the average values in the same rows indicate significant differences according to Duncan's multiple range tests after Bonferroni correctio (N = 4 plots).

[illegible]

Table 2. Average values and standard errors (±) of insect damage and fitness-related traits in weedy rice parent (W), transgene-positive (TP), and transgene-negative (TN) crop-weed hybrid progeny (F_4) in mixed planting under natural (above) vs. low (below) insect pressure.

doi:10.1371/journal.pone.0041220.t002

Trait	Plant material					
	TP vs. TN		TP vs. W		TN vs. W	
Natural insect						
Blasted tillers %	11.4±2.0	17.3±0.9**	9.6±2.1	13.0±2.3	12.9±1.4	18.1±3.0
Folded leaves %	2.9±1.7	5.0±3.0	4.2±1.3	5.4±1.2	15.6±2.7	10.3±3.7
Plant height (cm)	101.4±2.6	103.4±4.8	96.9±2.9	93.1±3.0	107.9±6.4	99.4±6.7
No. of tillers	11.0±1.0	9.9±0.8	15.2±1.2	12.0±1.0	11.8±0.5	9.8±1.0
No. of panicles	8.9±0.4	8.6±0.8	9.7±1.0	8.3±0.4	8.9±0.7	8.2±0.9
Seed set (%)	41.5±4.2	37.1±0.6	45.5±1.7	49.3±2.6	37.0±2.0	47.8±3.8*
1000-seed weight (g)	20.4±0.6	20.5±0.2	20.6±0.1	19.6±0.1*	21.1±0.6	19.1±0.3*
Low insect						
Blasted tillers %	5.4±0.3	5.8±0.4	5.3±1.5	6.5±1.9	5.7±2.2	8.2±1.7
Folded leaves %	0.4±0.3	0.8±0.3	0.6±0.1	0.7±0.2	2.3±0.4	0.7±0.3*
Plant height (cm)	112.3±5.0	116.7±3.4	104.5±4.7	98.3±3.4	111.4±2.8	104.1±1.6
No. of tillers	7.6±0.1	8.2±0.3	8.5±0.5	7.8±0.3	8.8±0.4	8.3±0.5
No. of panicles	6.8±0.4	7.2±0.3	6.8±0.4	6.2±0.3*	7.3±0.6	7.2±0.3
Seed set (%)	51.9±0.9	49.0±5.2	49.9±5.7	47.9±2.4	51.6±2.6	55.7±2.9
1000-seed weight (g)	22.3±0.8	21.9±0.5	21.9±0.7	19.3±0.4*	21.7±0.2	18.8±0.4**

Level of significance was calculated based on the paired *t*-test between TP and TN, TP and W, or TN and W (N = 4 plots).
*P<0.05;
**P<0.01.
[illegible]

Noticeably, the insect index was significantly lower in the mixed planting of transgene-positive populations than in the pure planting under natural insect pressure (Fig. 1), indicating that the mixed planting with transgene-positive plants will generally reduce the ambient insect pressure. However, such reduction of insect damage was not observed in the mixed planting of transgene-positive plants under low insect pressure.

Performance of F_4 and Weedy Parental Populations in Pure Planting

In general, transgene-positive F_4 populations with insect-resistance transgenes showed significantly less insect damage than transgene-negative F_4 populations and weedy parents, under natural insect pressure (Table 1). Transgene-positive F_4 populations had a better performance for most of the fitness-related traits than transgene-negative F_4 populations and weedy parents, under natural insect pressure (Table 1). For example, transgene-positive plants produced a significantly higher number of panicles (25% increase) and ratio of seed set (17% increase) than their transgene-negative plants (Table 1). Particularly, the number of seeds produced by transgene-positive plants was significantly higher than both transgene-negative plants (44% increase) and weedy parents (31% increase) (Fig. 2). Under low insect pressure, nearly all the fitness-related traits did not show significant differences between transgene-positive and transgene-negative populations (Table 1; Fig. 2) in both pure and mixed planting, suggesting no obvious fitness cost from the transgenes.

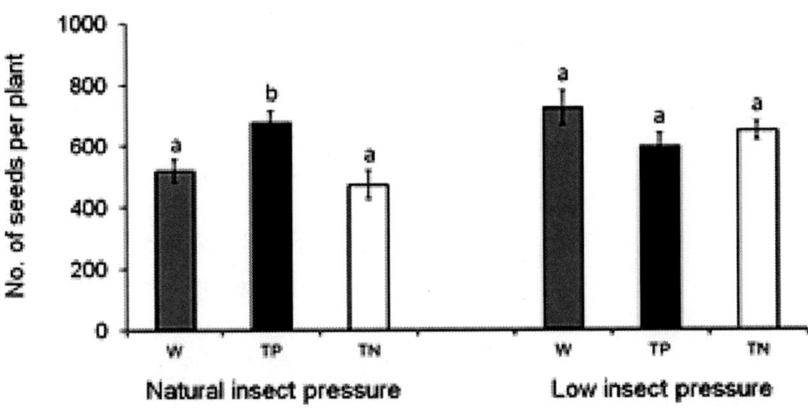

Figure 2. The average number of filled seeds per plant produced by weedy rice parent (W), transgene-positive (TP), and transgene-negative (TN) hybrid progeny under natural vs. low insect pressure in pure cultivation.

The comparison was made among weedy rice parent, transgene-positive, and transgene-negative hybrid progeny. Different letters above the columns indicate significant differences according to Duncan's multiple range tests after Bonferroni correction. Bars represent standard error. Levels of significance: P<0.05.

doi:10.1371/journal.pone.0041220.g002

Noticeably, transgene-negative F_4 populations showed significantly higher values of plant height and 1000-seed weight than their weedy rice parental populations under both natural and low insect pressures (Table 1). This indicated that crop-weed hybridization may bring benefit through the enhanced performance of some traits of hybrids and hybrid lineages in later generations.

Performance of F_4 and Weedy Parental Populations in Mixed Planting

In contrast to the pure planting, transgene-positive and transgene-positive F_4 populations showed apparently less pronounced differences in insect damage in mixed planting than in pure planting under natural insect pressure (Table 2). The same trend was observed between transgene-positive F_4 and weedy parental populations (Table 2) in mixed planting. As a result, transgene-positive F_4 populations did not show significant differences for nearly all the fitness-related traits, compared with transgene-negative F_4 and weedy parental populations under natural insect pressure (Table 2). These results indicated that the mixed-planting plots with transgene-positive plants would result in a reduced insect damage to transgene-negative and weedy rice plants, and consequently reduce the differences in fitness-related traits between transgene-positive and transgene-negative plants.

Similarly, under low insect pressure, nearly all the fitness-related traits did not show significant differences among transgene-positive F_4, transgene-negative F_4 and weedy parental populations (Table 2; Fig. 3). However, it is worthy of notice that the 1000-seed weight of transgene-positive and transgene-negative F_4 plants was significantly higher than that of weedy parental plants, suggesting again the benefit of hybridization brought to hybrid progeny compared with their weedy parents in terms of seed weight, regardless of the presence of transgenes.

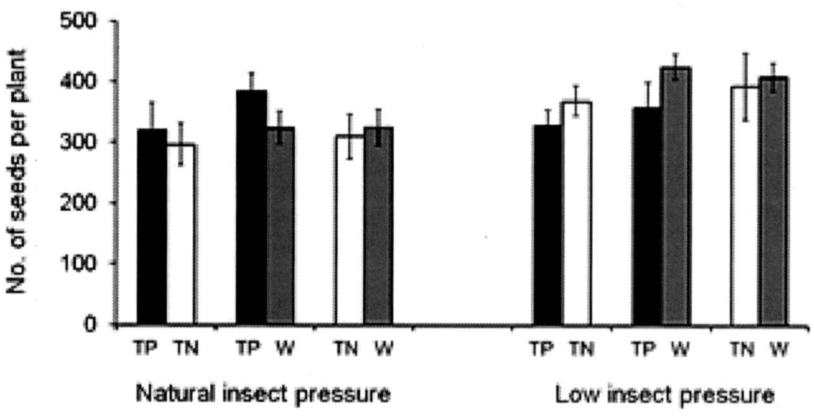

Figure 3. The average number of filled seeds per plant produced by weedy rice parent, transgene-positive (W), and transgene-negative (TP) hybrid progeny under natural vs.

Low insect pressure in mixed cultivation

The comparison was made between transgene-positive and transgene-negative, transgene-positive and weedy rice parent, or transgene-negative and weedy rice parent, based on the paired *t*-tests. Bars represent standard error. Levels of significance: P<0.05.

doi:10.1371/journal.pone.0041220.g003

DISCUSSION

The key point of assessing potential ecological risks caused by an insect-resistant transgene introgressed to weedy populations is to analyze the increased fitness of transgenic plants in comparison with their nontransgenic counterparts derived from the same crop-weed hybridization under controlled environmental conditions [11], [22], [24], [25], [31]. Our data from this study indicated significant differences in insect pressure (as measured by the insect index,Fig. 1) between the two rice planting environments with natural *vs.* low insect pressure, achieved by artificial spray of insecticides. The insect pressure in both pure and mixed planting plots of weedy rice parents and transgene-negative F_4 populations was significantly higher under the natural-insect condition than that under the controlled low-insect condition. These results indicate the effectiveness of insect control in our experiment. In addition, as indicated by the insect damage on weedy parents and transgene-negative plants, significantly higher insect index was also detected in pure planting plots than in the mixed-planting plots of nontransgenic plants (including parents) with transgenic plants under natural insect. In contrast, the mixed-planting plots with transgenic plants did not show such significant differences in insect index under low insect pressure. All together, these results demonstrate that the differential environment allowed us to analyze the fitness effect of insect-resistance transgenes in weedy rice plants against their nontransgenic counterparts, and that the mixed planting of transgenic plants in a field plot can significantly reduce the ambient insect pressure.

In pure planting, transgene-positive crop-weed F_4 plants showed significantly greater values of fecundity-related traits, such as the number of panicles and well-filled seeds and the ratio of seed set than their transgene-negative plants under natural insect pressure. Particularly, the number of panicles and filled seeds of transgene-positive F_4 plants showed significantly higher values than the weedy rice parents. However, such significant differences between the transgenic plants and nontransgenic plants were not detected in these traits under the low insect pressure. The increased fecundity of transgenic crop-weed F_4 plants is most likely brought by the introgression of insect-resistance transgenes that significantly reduced insect attack to the plants. The increased fecundity in the transgenic F_4 plants demonstrates that the introgressed insect-resistant transgenes (*Bt/CpTI*) are still effective on controlling rice target herbivores in the advanced generation of crop-weed hybrid populations. This result is consistent with our previous study in which the F_2–F_3 lineages derived from the crop-weed hybrids with the same transgenic event showed effective control of insect damages to the hybrid lineage compared with both transgene-negative hybrid lineages and the weedy parents, under a high level of insect pressure [25], [29]. Many previous studies on crop-weed and crop-wild hybrid lineages also demonstrated that insect-resistance transgenes can dramatically reduce the target herbivores and increase the fecundity in hybrids and their advanced generation of hybrid populations [23], [25], [29]. These findings are consistent with our results in this study. It seems possible to make a conclusion from the field experimental data of above studies including ours that the introgressed insect-resistance transgenes will maintain their strong ability to control herbivores and increase fecundity of transgenic populations derived from crop-weed or crop-wild transgene introgression. The transgenic plants may have a strong ability to compete with their nontransgenic counterparts including nontransgenic hybrid and parental populations[23], [25], [28].

However, it is necessary to point out that in the mixed-planting plots the fecundity traits (particularly seed production) of the transgene-positive plants did not show significantly increased values, compared to the transgene-negative plants and weedy rice parents under natural insect pressure. This result suggests apparent losses of fitness benefit that should have been brought by the insect-resistance transgenes in mixed planting under natural insect pressure. The similar phenomenon was also observed in a number of studies in which the fitness benefit and cost of insect-resistant GE rice lines and their nonGE parental lines was analyzed under natural and low insect pressure, respectively [22], [24]. Associated with the data of insect index from this study where significantly lower insect pressure was recorded in the mixed planting plots than in the pure planting plots under the same environmental condition of natural insect (Fig. 1), we think that the loss of fitness benefit to insect-resistant transgenic crop-weed hybrid populations in mixed planting is most likely owing to the considerably reduced insect pressure. In other words, the generally low ambient insect pressure in the experimental field caused by the mixture of insect-resistant transgenic plants will significantly reduce the potential fitness advantages that should have been brought by insect-resistance transgenes. As a result, the expected long-term persistence and rapid spread of insect-resistance transgenes in weedy rice populations caused considerably increased fitness advantages following transgene flow from a GE rice variety may not happen in the realistic situation if such fitness advantages are extremely limited. Under the actual situation where an insect-resistant GE rice variety is cultivated, the ambient insect pressure in an extensive field area should be reduced to a much greater extent than that in our experimental plots. The spread of transgenes in weedy rice populations might be considerably limited owning to the negligible fitness increase in insect-resistance GE rice fields where the target insect pressure is significantly reduced, although the issue of transgene flow from an insect-resistant GE rice variety to its coexisting weedy rice populations should not be neglected.

Our finding has its important implications for the risk assessment of transgene flow from insect-resistant GE crop to its wild relatives, and to the conspecific weedy populations in particular. Given a determined frequency of transgene flow, the magnitude of potential environmental risks should largely depend on the fitness effect of a transgene. It is generally recognized that fitness effect of an insect-resistance transgene is determined by the ambient insect pressure in the environment where wild/weedy populations occur. Therefore, the assessment of environmental risks caused by transgene flow from an insect-resistant GE crop to wild populations should first consider the pressure of target insects in the concerned environment. This principle may also applied to the risk assessment of transgene flow to wild populations, considering that insect pressure in natural habitats is significantly lower than in agriculture habitats [32]. For the crop conspecific weedy populations co-occurring with a crop (and GE crop), crop-to-weed transgene flow cannot be avoided. The expected environmental impact from transgene flow from insect-resistant GE crop could be large because hypothetically insect-resistance transgenes will bring fitness benefit to the weedy populations. However, the fitness advantages might be limited due to the fact that weedy plants will be surrounded by insect-resistant plants in a GE crop field, reducing the ambient pressure of target herbivores significantly. Consequently, the spread of the transgenes in weedy populations would be limited. In reality, the extensive commercial cultivation of an insect-resistant GE crop will largely reduce target herbivores in a GE deployed area [33]. Under such a circumstance, the environmental impacts caused by the crop-to-weed transgene flow from an insect-resistant GE crop will also be limited.

AUTHOR CONTRIBUTIONS

Conceived and designed the experiments: BRL XY. Performed the experiments: XY JS. Analyzed the data: BRL XY FW. Contributed reagents/materials/analysis tools: FW JS. Wrote the paper: XY BRL.

REFERENCES

1. Andow DA, Zwahlen C (2006) Assessing environmental risks of transgenic plants. Ecol Lett 9: 196–214.
2. Conner AJ, Glare TR, Nap JP (2003) The release of genetically modified crops into the environment - Part II. Overview of ecological risk assessment. Plant J 33: 19–46.
3. Snow AA, Palma PM (1997) Commercialization of transgenic plants: Potential ecological risks. Bioscience 47: 86–96.
4. Haygood R, Ives AR, Andow DA (2003) Consequences of recurrent gene flow from crops to wild relatives. P Roy Soc B-Biol Sci 270: 1879–1886.
5. Lu BR, Yang C (2009) Gene flow from genetically modified rice to its wild relatives: Assessing potential ecological consequences.

Biotechnol Adv 27: 1083–1091.

6. Stewart CN, Halfhill MD, Warwick SI (2003) Transgene introgression from genetically modified crops to their wild relatives. Nat Rev Genet 4: 806–817.
7. Ellstrand NC, Prentice HC, Hancock JF (1999) Gene flow and introgression from domesticated plants into their wild relatives. Annu Rev Ecol Evol S 30: 539–563.
8. Ellstrand NC (2003) Current knowledge of gene flow in plants: implications for transgene flow. Philos T R Soc B 358: 1163–1170.
9. Raybould AF, Gray AJ (1993) Genetically modified crops and hybridization with wild relatives: A UK Perspective. J Appl Ecol 30: 199–219.
10. Chen LJ, Lee DS, Song ZP, Suh HS, Lu BR (2004) Gene flow from cultivated rice (*Oryza sativa*) to its weedy and wild relatives. Ann Bot 93: 67–73.
11. Lu B-R, Snow AA (2005) Gene flow from genetically modified rice and its environmental consequences. BioScience 559: 669–678.
12. Gealy DR, Mitten DH, Rutger JN (2003) Gene flow between red rice (*Oryza sativa*) and herbicide-resistant rice (*O. sativa*): implications for weed management. Weed Technol 17: 627–45.
13. Warwick SI, Légère A, Simard MJ, James T (2008) Do escaped transgenes persist in nature? The case of an herbicide resistance transgene in a weedy *Brassica rapa*population. Mol Ecol 17: 1387–1395.
14. Hails RS, Morley K (2005) Genes invading new populations: a risk assessment perspective. Trend Ecol Evol 20: 245–252.
15. FAO (2004) The state of food and agriculture 2003–2004. Agricultural biotechnology: meeting the needs of the poor? (Accessed on April 22, 2012).
16. Huang JK, Hu RF, Rozelle S, Pray C (2005) Insect-Resistant GM Rice in Farmers' Fields: Assessing Productivity and Health Effects in China. Science 308: 688–690.
17. MOA (Ministry of Agriculture, China) (2009) The second batch of genetically modified organisms that are granted biosafety certificates in 2009. (in Chinese). (Accessed on April 22, 2012).
18. Delouche JC, Burgos NR, Gealy DR, Zorilla-San MG, Labrada R, et al. (2007) Weedy rices: origin, biology, ecology and control.

FAO of the United Nations, Rome. 144 p. (Accessed on April 22, 2012).

19. Londo JP, Schaal BA (2007) Origins and population genetics of weedy red rice in the USA. Mol Ecol 16: 4523–4535.
20. Harlan JR, De Wet J (1971) Toward a rational classification of cultivated plants. Taxon 20: 509–517.
21. Cohen MB, Arpaia S, Lan LP, Chau LM, Snow AA (2008) Shared flowering phenology, insect pests, and pathogens among wild, weedy, and cultivated rice in the Mekong Delta, Vietnam: implications for transgenic rice. Environ Biosafety Res 7: 73–85.
22. Chen LY, Snow AA, Wang F, Lu BR (2006) Effects of insect-resistance transgenes on fecundity in rice (*Oryza sativa*, Poaceae): A test for underlying costs. Am J Bot 93: 94–101.
23. Snow AA, Pilson D, Rieseberg LH, Paulsen MJ, Pleskac N, et al. (2003) A Bt transgene reduces herbivory and enhances fecundity in wild sunflowers. Ecol Appl 13: 279–286.
24. Xia H, Chen LY, Wang F, Lu BR (2010) Yield benefit and underlying cost of insect-resistance transgenic rice: Implication in breeding and deploying transgenic crops. Field Crop Res 118: 215–220.
25. Yang X, Xia H, Wang W, Wang F, Su J, et al. (2011) Transgenes for insect resistance reduce herbivory and enhance fecundity in advanced generations of crop-weed hybrids of rice. Evol Appl 4: 672–684.
26. Halfhill MD, Sutherland JP, Moon HS, Poppy GM, Warwick SI, et al. (2005) Growth, productivity, and competitiveness of introgressed weedy *Brassica rapa* hybrids selected for the presence of *Bt cry1Ac* and *gfp* transgenes. Mol Ecol 14: 3177–3189.
27. Mason P, Braun L, Warwick SI, Zhu B, Stewart CN (2003) Transgenic *Bt*-producing*Brassica napus*: *Plutella xylostella* selection pressure and fitness of weedy relatives. Environ Biosafety Res 2: 263–276.
28. Vacher C, Weis AE, Hermann D, Kossler T, Young C, et al. (2004) Impact of ecological factors on the initial invasion of Bt transgenes into wild populations of birdseed rape (*Brassica rapa*). Theor Appl Genet 109: 806–814.

29. Cao QJ, Xia H, Yang X, Lu BR (2009) Performance of hybrids between weedy rice and insect-resistant transgenic rice under field experiments: implication for environmental biosafety assessment. J Integr Plant Biol 51: 1138–1148.

30. Rong J, Song ZP, Su J, Xia H, Lu BR, et al. (2005) Low frequency of transgene flow from*Bt/CpTI* rice to its nontransgenic counterparts planted at close spacing. New Phytol 168: 559–566.

31. Jenczewski E, Ronfort J, Chèvre AM (2002) Crop-to-wild gene flow, introgression and possible fitness effects of transgenes. Environ Biosaf Reserv 2: 9–24.

32. Chen YH, Welter SC (2002) Abundance of a native moth *Homoeosoma electellum*(Lepidoptera: Pyralidae) and activity of indigenous parasitoids in native and agricultural sunflower habitats. Environ Entomol 31: 626–636.

33. Wu KM, Lu YH, Feng HQ, Jiang YY, Zhao JZ (2008) Suppression of cotton bollworm in multiple crops in China in areas with Bt toxin-containing cotton. Science 321: 1676–1678.

Chapter 4

RESISTANCE EVOLUTION TO BT CROPS: PREDISPERSAL MATING OF EUROPEAN CORN BORERS

[1]Ambroise Dalecky, [2] Sergine Ponsard, [1,2]Richard I Bailey, [2]Céline Pélissier, [1]Denis Bourguet

[1]Institut National de la Recherche Agronomique (INRA), UMR Centre de Biologie et de Gestion des Populations (CBGP), Campus International de Baillarguet, Montferrier-sur-Lez, France

[2]Laboratoire Dynamique de la Biodiversité, CNRS-Université P. Sabatier–Toulouse III, Toulouse, France

ABSTRACT

Over the past decade, the high-dose refuge (HDR) strategy, aimed at delaying the evolution of pest resistance to Bacillus thuringiensis (Bt) toxins produced by transgenic crops, became mandatory in the United States and is being discussed for Europe. However, precopulatory dispersal and the mating rate between resident and immigrant individuals, two features influencing the efficiency of this strategy, have seldom been quantified in pests targeted by these toxins. We combined mark-recapture and biogeochemical marking over three breeding seasons to quantify these features directly in natural populations of Ostrinia nubilalis, a major lepidopteran corn pest. At the local scale, resident females mated regardless of males having dispersed beforehand or not, as assumed in the HDR strategy. Accordingly, 0–67% of resident females mating before dispersal did so with resident males, this percentage depending on the local proportion of resident males (0% to 67.2%). However, resident males rarely mated with immigrant females (which mostly arrived mated), the fraction of females mating before dispersal was variable and sometimes substantial (4.8% to 56.8%), and there was no evidence for male premating dispersal being higher. Hence, O. nubilalis probably mates at a more restricted spatial scale than previously assumed, a feature that may decrease the efficiency of the HDR strategy under certain circumstances, depending for example on crop rotation practices.

INTRODUCTION

Right after emergence and just before mating, adult insects face a crucial dilemma that The Clash [1] celebrated in a famous song that could be slightly rephrased as: "Should I mate or should I go now? If I mate there will be trouble. But if I go it might be double." In other words, fitness is affected by two important adult "decisions" – mating and dispersal – and by the order in which they are performed. Mating before dispersal may increase the risk of consanguinity, but dispersal before mating may increase the risk of not finding a sexual partner. Therefore, the timing between mating and dispersal is likely to vary with species and environmental conditions [2]: in the absence of a universal optimal strategy, no general prediction can be made and each species of interest must be studied on its own.

Nevertheless, the timing between mating and dispersal can be of great practical importance: notably, it may influence the efficacy of strategies intended to drive the microevolution of agricultural pest species – e.g., the evolution of their resistance to control agents such as pesticides – by managing agricultural landscapes [3]. The "high-dose refuge" (HDR) strategy [4] is one such strategy. It is aimed at delaying or preventing the evolution of resistance in target pest populations against Bacillus thuringiensis (Bt) toxins produced by transgenic Bt crops [5]. The underlying principle is that, in a patchy environment of treated and untreated areas, high gene flow between patches with different selection pressure (here, between Bt crop fields andBt-free refuges) should limit local adaptation [6] (here, the selection of resistance alleles). Over the past decade, an HDR-based management of agricultural land became mandatory for severalBt crops in the United States [5] and is being discussed for Europe [7].

Population genetics studies conducted on the main pests targeted by the Bt crops usually revealed no departure from patterns expected at Hardy-Weinberg equilibrium and suggested a high level of gene flow over a broader spatial scale than that of the patches, a necessary condition for the success of the HDR strategy (e.g., [8–10]). However, tests for departure from Hardy-Weinberg equilibrium are quite conservative, estimates of the spatial scale at which gene flow occurs based on such methods are coarse, and, more fundamentally, the same genetic structure patterns can also be generated by a number of alternative processes [11].

In particular, population genetics studies provide no information on whether adults mate mostly before or mostly after dispersal. However, high predispersal mating can reduce the efficacy of the HDR strategy. Indeed, the goal of this strategy is to purge each generation of as many as possible of its "r" alleles (alleles conferring resistance to the Bt toxin) to counterbalance, as much as possible, the increase of their frequency resulting from the selection pressure exerted by Bt crops. The r alleles are carried by either heterozygous (rS) or homozyous (rr) individuals. Resistance to Bt toxins being generally recessive [12], only the fraction of r alleles carried byrS individuals among the offspring of any given generation is thus available for purging. Adults emerging from a Bt field are likely to be mostly resistant homozygotes (rr), whereas adults emerging from refuges are mostly susceptible homozygotes (SS). If most mating takes place before dispersal, the proportion of heterozygotes in the offspring is expected to be low. Other things being equal (but see, e.g., [13] and Discussion), the higher the predispersal mating rate, the lower is the expected rS/(rS + rr) ratio in the offspring and the lower is the expected success of the HDR strategy.

Classical mark-recapture studies have been performed on a

number of agricultural pests (e.g., [14–16]), mostly with the aim of estimating dispersal distances. Although they provide more precise estimates of such distances than population genetics studies, only very few of them examined the timing between mating and dispersal. This relationship has, to our knowledge, seldom and incompletely been quantified directly in natural populations of insect pests targeted by Bt crops (but see [16–18]). First, it is often difficult to detect individuals in the field at the very moment they mate, and even more difficult to appreciate their mating success in terms of sperm transfer. Second, and most important, the origin—local or external—of individuals mating in a specific site can generally not be ascertained.

The present study deals with the European corn borer (ECB), Ostrinia nubilalis Hübner (Lepidoptera: Crambidae), one of the major pests of corn (Zea mays L.), and one of the main targets of the Cryab and Cry1F Bt toxins produced by transgenic Bt corn [19]. Since the 2000 growing season, US corn belt growers planting Bt corn outside and inside Bt cotton growing areas must assign 20% and 50%, respectively, of their corn acreage to refuges, and must set these refuges less than 0.5 mile (approximately 800 m)—and preferably less than 0.25 mile (approximately 400 m)—away from Bt cornfields [20]. When these requirements were made, empirical data about ECB dispersal were very scarce [20, 21]. Studies published since then [22–24] confirmed that dispersing ECB moths are able to move over far more than a few hundred meters, but the proportion of individuals actually engaging in dispersal, the probability that they do so before versus after mating, and the probability that immigrants and residents mate randomly remained open questions.

We used a combination of color-mark, release and recapture experiments and of biogeochemical marking [25] over three breeding seasons (one in 2004 and two in 2005) to quantify, directly in natural populations, the percentage of ECB individuals staying long enough to mate in their natal cornfield and its herbaceous border and the percentage of mating actually occurring between those individuals when mixed with dispersing individuals coming in from surrounding cornfields. Our results show that immigrant males, once present locally, have the same probability of mating with resident females as resident males have, while limited evidence suggests that immigrant females mate less readily with local males than do resident females.

They also show that a variable and sometimes substantial proportion of ECB females mate before they engage in any dispersal, with no evidence for any sex-related difference. This might decrease the efficiency of the HDR strategy in certain situations.

RESULTS

Experiment 1

Experiment 1 aimed at estimating the probability of young (less than 24 h old) individuals located in a cornfield to be found in the closest field border two nights later and at estimating the probability that such young individuals mate assortatively rather than with immigrants coming from other cornfields. Over 17 sessions spread over 3 wk of the June 2004 breeding season, we released a total of 8,788 virgin moths into four different cornfields (Table 1). These fields were planted with soybean (Glycine max L.) or wheat (Triticum sativum L.) during the previous (2003) season (two plants that suffer no or very little infestation by the ECB [26, 27]), thus ensuring that virtually no ECB pupae from the previous season were present locally. In this respect, such fields mimicked fields planted with Bt corn during the previous season. The closest fields planted with corn in 2003—mimicking the refuges—were located at a distance of at least 400 m. ECB larvae having overwintered in these 2003 cornfields were the closest, and hence the most likely, sources of wild moths occurring at the study sites in 2004.

Table 1. Experiment 1

Site	Date of Release	n Released		Percent Recaptured	
		Males	Females	Males	Females
1	01 June	268	233	32.8	20.2
	07 June	300	300	2.0	6.3
	11 June	300	202	1.3	6.4
	16 June	300	273	1.7	1.8
	23 June	200	140	0.5	2.1
2	31 May	300	268	5.3	9.0
	06 June	226	244	1.8	0.4
	10 June	300	221	0.3	0
	14 June	296	256	0	0.4
3	30 May	281	257	15.7	3.9
	05 June	284	196	2.5	1.5
	09 June	300	264	2.3	1.5
	13 June	246	300	0.8	0.7
4	29 May	257	193	5.1	2.1
	02 June	275	217	6.6	7.8
	08 June	300	215	0.3	0
	12 June	300	276	0.7	0.7

Number of *O. nubilalis* males and females released and percentage recaptured for each release-recapture session.

DOI: 10.1371/journal.pbio.0040181.t001

doi:10.1371/journal.pbio.0040181.t001

At each site and for each release session, 340 to 600 marked moths that had emerged less than 24 h earlier were released into the cornfield, between the seventh and eighth rows, planted parallel to the field border (Table 1). Released moths were marked twice: they were reared as larvae on a wheat (a C3 plant) diet, which caused their tissues to contain a lower ratio of13C/12C carbon isotopes than wild individuals fed on corn (a C4 plant) [27–31] without altering their longevity (Table S1) or propensity to mate (Table S2), and they were marked with a colored ink spot on the dorsal thorax. Thirty-six hours (2 nights and 1 day) after release, we recaptured a total of 374 marked moths in the herbaceous borders of the field, along with a total of 5,504 unmarked individuals (Table 2).

Table 2. Experiment 1

Site	Date of Recapture	Males		Females		Marked Females		
		n	Percent Marked	n	Percent Marked	n	Percent Mated	Percent Mated by Marked Males
1	01 June	515	27.9	243	15.5	47	95.7	13
	09 June	386	3.1	453	6.8	14	100	5
	13 June	347	8.5	1308	0.8	11	100	18
	16 June	400	1.3	521	1.5	7	100	0
	25 June	985	1.2	128	2.1	3	100	0
2	02 June	37	43.2	25	96.0	24	100	58
	08 June	10	40.0	3	33.3	1	0	—
	12 June	10	10.0	12	0.0	0	—	—
	16 June	4	0.0	2	50.0	1	100	0
3	01 June	64	67.2	13	85.5	18	93.8	42
	07 June	82	8.5	58	5.2	3	100	0
	11 June	79	10.1	50	8.0	4	100	0
	15 June	29	3.4	83	3.6	3	100	0
4	31 May	32	25.0	33	9.1	3	66.7	50
	04 June	98	18.4	86	20.9	18	100	33
	10 June	41	2.4	67	0.0	0	—	—
	14 June	29	3.4	86	2.3	2	100	50

Percentage of marked individuals among [illegible] males and females captured [illegible] after release into a [illegible] section situated between the seventh and eighth rows of a cornfield, in the closest [illegible] section of the parallel herbaceous field border, and percentage of marked spermatophores carried by marked females.

DOI: 10.1371/journal.pbio.0040181.t002

doi:10.1371/journal.pbio.0040181.t002

The proportion of marked individuals among all moths caught varied widely between sites and sessions, as did the proportion of individuals released during a given session that were recaptured (Tables 1 and 2). The percentage recaptured averaged (±1 SE) 4.3 ± 1.6% and ranged from 0.2% to 26.9%. This proportion was not significantly different between sexes (paired t-test: $t = 0.73$, $df = 16$, $p = 0.762$). Approximately 97% of the 155 females we recaptured were mated (Table 2). To determine whether they had mated with wild or with released males, the spermatophore (or the most recent spermatophore, in cases of multiple matings) dissected from the bursa copulatrix of their genital duct were subjected to stable carbon isotope analysis [27, 31]. As expected, the δ13C values of the 150 spermatophores showed a bimodal distribution. Based on previous results [27], spermatophores with δ13C values of −31‰ to −22‰ were assigned to released males reared on wheat, whereas those with δ13C values of −20‰ to −9‰ were assigned to wild (corn-fed) males originating from surrounding cornfields. A small number of spermatophores (n = 5, less than 4%) with intermediate δ13C values (−22‰ to −20‰) were considered of uncertain origin and excluded from our analysis. This biogeochemical method therefore resulted in the assignment of greater than 96% of the mated females' sexual partners to either the wild or the released males' pool.

Marked and recaptured females had mated with both types of males. Over all sites, the proportion of released males among their partners was 19.4%. It ranged from 0% to 67%, differing considerably over the 17 release-recapture sessions (Table 2), and it was positively correlated with the ratio of the number of released males over the total number of males captured in the herbaceous border during the corresponding session (Spearman's rank correlation coefficient rS = 0.622, $p < 0.05$, n = 14; Table 2 and Figure 1). In accordance with this result, the maximum-likelihood estimate of the proportion of released females mating assortatively with released rather than wild males was very low (mode = 0.04), and its 95% credibility interval included 0 (Figure 2).

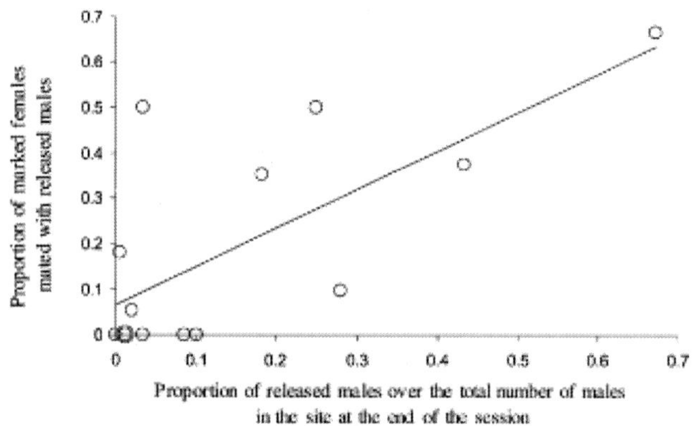

Figure 1. Frequency of Crossing among Resident Moths as a Function of the Local Abundance of Resident Males Experiment 1. Correlation between the proportion of marked among spermatophores dissected from marked O. nubilalis females recaptured during a given session and the proportion of marked males among the total number of males captured during recapture during the same session (rS = 0.622, $p < 0.05$, n = 14).

doi:10.1371/journal.pbio.0040181.g001

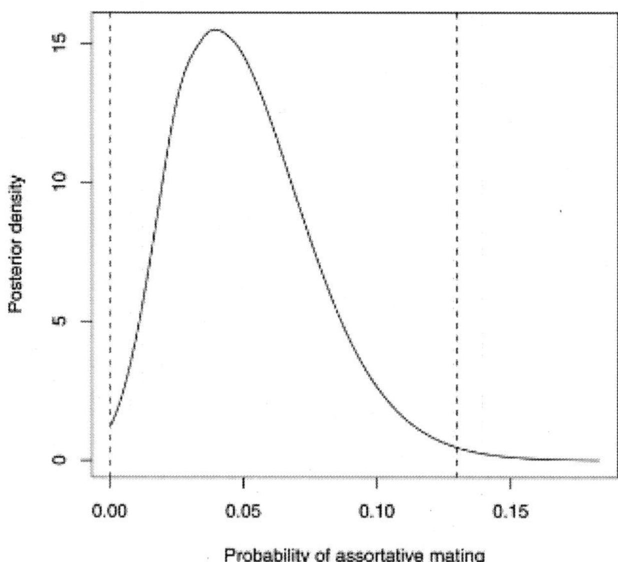

Figure 2. Posterior Density of the Probability of Assortative Mating Experiment 1. Posterior distribution of the probability of assortative mating (pAij) among released O. nubilalis moths, as estimated using a Bayesian Markov chain Monte Carlo approach (see text for details). The dashed lines indicate the 95% credibility interval (0.00 to 0.13).

doi:10.1371/journal.pbio.0040181.g002

We also determined the δ13C values of the spermatophores dissected from the bursa copulatrix of 79 wild females caught during sessions displaying the highest released/wild male ratios. We found that all but three wild females had mated with wild males and that the overall proportion of females mated with wild males was thereby significantly higher among wild females (96.2%) than among recaptured females (80.6%; Fisher's exact test, $p = 0.0003$). Wild females also carried more numerous (Fisher's global test, $\chi2 = 19.13$, $df = 8$, $p = 0.014$) and older (Fisher's global test, $\chi2 = 71.24$, $df = 8$, $p < 0.0001$) spermatophores than recaptured females did (Tables S3 and S4). Furthermore, wild females captured in the study sites carried more numerous (Fisher›s global test, $\chi2 = 16.15$, $df = 8$, $p = 0.040$) and older (Fisher's global test, $\chi2= 72.00$, $df = 8$, $p < 0.0001$) spermatophores than did wild females caught in the border of nearby cornfields planted with corn the previous year; i.e., the closest and therefore the most likely sources of most wild moths found in the study sites (Tables S3 and S4).

Although very informative, Experiment 1 had two drawbacks. First, although the moths we released were very young (less than 24 h postemergence) and although the proportion of recaptured moths did not vary with age upon release (less than 12 h or 12 to 24 h; seeMaterials and Methods), we could not entirely discard the possibility that, in natural conditions, some dispersal occurs during the very first hours of adult life and that the moths we released had already lost this propensity to disperse. Second, Experiment 1 most probably underestimated the proportion of moths mating in the vicinity of their natal field. Indeed, the herbaceous border section screened during any recapture session was only a small part of the entire field border. Moreover, given that both male and female ECBs can mate a few hours after emergence [32], some released moths might have mated locally and left the site before the recapture session. Experiments 2 and 3, performed the following year, were designed to fill these two gaps.

Experiment 2

Experiment 2 was designed to characterize moth movement right after emergence, directly in field conditions. In June 2005, we monitored the time of emergence of a batch of ECB pupae placed directly in the field. A total of 221 emergences were recorded. There were two emergence peaks: 51.6% of pupae emerged in the morning (07:00 A.M. to 02:00 P.M., mostly [more than 90%] between 09:00 A.M. and 11:00 A.M.), and 48.4% emerged in the evening (04:00P.M. to 07:00 A.M., mostly [more than 90%] between 07:00 P.M. and 03:00 A.M.).

However, regardless of their sex and period of emergence, more than two thirds of the moths performed their first flight between 09:00 A.M. and 02:00 P.M. during the first 24 h of their adult life. The time of day of the first flight did not vary with period of emergence ($F_{1, 165} = 0.452$, $p = 0.502$), and varied only slightly with sex (females: 02:04 P.M. ± 27 min, range: 08:55 A.M. to 10:45 P.M., n = 68; males: 00:48 P.M. ± 17 min, range: 08:50 A.M. to 10:24 P.M., n = 101; $F_{1,165}= 4.347$, $p = 0.039$), with no significant interaction between sex and period of emergence ($F_{1,165} = 0.595$, $p = 0.442$). This confirms quantitatively our observations that individuals emerging in the evening typically move relatively little during their first night, and subsequently move essentially like individuals emerging the next morning.

The average distance covered during the first flight after emergence also did not vary with period of emergence ($F_{1,120} = 2.309$, $p = 0.131$) or sex ($F_{1,120} = 0.870$, $p = 0.353$; effect of the interaction between sex and period of emergence: $F_{1,120} = 0.017$, $p = 0.896$). Over all moths, the average distance (±1 SE) covered during this first flight was short (0.95 ± 0.13 m, range: 0.10 to 10.20 m, n = 124) and not significantly different from that of subsequent daytime movements (0.94 ± 0.25 m, range: 0.10 to 10.00 m, n = 52; $t = -0.034$, df = 174, $p = 0.973$). The average time lag (±1 SE) between successive flights was 26.2 ± 1.1 min, with a 95% confidence interval of 19.3 to 40.5 min. The frequency of these flights, recorded during 136 observation sessions of 10 min each, did not vary with period of emergence ($F_{1,123} = 0.278$, $p= 0.599$) or sex ($F_{1,123} = 1.067$, $p = 0.304$; effect of the interaction between sex and period of emergence: $F_{1,123} = 0.264$, $p = 0.608$). Hence, even a moth performing its first flight earlier than most individuals (say, at 06:00 A.M.) and moving always in the same direction would have a 95% probability of being located less than 67 m from its emergence point at 10:00 P.M. (average ± 1 SE: 34.5 ± 16.4 m, 95% confidence interval: 3.1 to 66.9 m).

Experiment 3

Experiment 3 aimed at examining the proportion of virgin ECB females mating locally, i.e., within 50 m of the cornstalk on which they are settled at dusk, over one night. As shown in Experiment 2, this stalk is situated less than 67 m from the moth›s place of emergence in 95% cases (34.5 m on average). To this effect, we released a total of 747 less-than-24-h-old virgin females by settling them on individual cornstalks, during both the first (June) and the second (August) 2005 breeding season. In June, the virgin ECBs were obtained from an outbred laboratory strain and color-marked before release. In August, the virgin ECBs were obtained as offspring of wild moths collected during the first flight, and they were not color-marked before release, for us to be able to exclude any effect of ink or laboratory selection on our results.

During the first flight season, the average (± 1 SE) proportion of females recaptured within the release section was 36.7 ± 8.6% (range: 16.7% to 62.2%; n = 5 release sessions; Table 3). During the second flight, this proportion was 15.6 ± 2.5% (range: 7.6% to 27.8%; n = 8 release sessions; Table 3). Both proportions were significantly different from each other (Mann-Whitney test, p = 0.040) (Table 3). This difference may result partly from moths being more easily overlooked in August (corn height: approximately 2.5 m) than in June (corn height: approximately 0.5 m), and partly from variations in the propensity to disperse due to environmental factors such as temperature.

Table 3. Experiment 3

Release Session	n	Released Females			
		Percent Recaptured			Percent Not Recaptured
		Mated Very Locally	Mated Locally	Recaptured, Unmated	
11 June	30	26.6	0	10.0	63.0
12 June	67	44.8	3.3	4.5	47.8
13 June	44	18.2	11.4	2.3	68.2
14 June	37	45.9	10.8	16.2	27.0
15 June	30	13.3	3.3	3.3	80.0
Total June	198	28.8	6.1	7.1	58.1
8 August	42	19.0	—	2.4	78.6
9 August (I)[a]	73	17.8	—	2.7	79.5
9 August (NI)[a]	62	8.1	—	0	91.9
10 August	97	6.2	—	5.2	88.7
12 August	145	4.8	—	2.8	92.4
13 August	56	7.1	—	5.4	87.5
14 August	38	5.3	—	10.5	84.2
16 August	36	25.0	—	2.8	72.2
Total August	549	9.8	—	3.6	86.5

[a]These two release sessions were performed on the same night in an irrigated (I) and a nonirrigated (NI) part of the field.

Number of O. nubilalis females released and percentage of females that mated very locally (i.e., within the release section), that mated locally (within 50 m of the stalk of release), recaptured within the release section after one night but not mated, or not recaptured (i.e., lost).

DOI: 10.1371/journal.pbio.0040181.t003

doi:10.1371/journal.pbio.0040181.t003

During the first flight season, taking into account only females recaptured within the release section itself—often on the very stalk where they had been released—the proportion of mated females over the total number of released females averaged (±1 SE) 28.4 ± 7.0% (range: 13.3% to 45.9%; n = 5 release sessions; Table 3). If we include females recaptured within 50 m from the stalk of release, this estimate rises to an average of 34.1 ± 7.8% (range: 16.7% to 56.8%; n= 5 release sessions; Table 3). During the second flight season, the proportion of females observed mating within the release section itself averaged (±1 SE) 11.7 ± 2.7% (range: 4.8% to 25.0%; n = 8 release sessions; Table 3). Over both breeding seasons, the average proportion (±1 SE) of females mating locally was 18.1 ± 3.8% (n = 13 release sessions; Table

3).

Experiment 3 was also attempted on virgin less-than-24-h-old males. Unfortunately, they proved to be too agitated to be settled successfully on a cornstalk at dusk. They were usually no longer present within 50 m of their initial location after approximately 30 min.

DISCUSSION

Since the HDR strategy has been chosen in the United States with the aim of delaying pest adaptation to transgenic insecticidal crops, a number of its potential limitations have been examined (see [5, 7, 33]). One of its main technical problems is that significant preferential mating among resistant individuals could decrease its efficiency [4]. However, despite a large modeling effort [4, 13, 34–41] and despite calls for data on how far adults move before they mate in order to determine the most suitable management strategy (e.g., [33]), empirical studies addressing this question remained scarce.

In the case of the ECB, a major target of the Bt toxins produced by transgenic insecticidal corn, the few studies on dispersal that have been published so far [22–24] provide estimates on the distance moved by this pest but little indication on the level of mating between immigrant and resident individuals or on the timing between dispersal and mating. These features are important because they directly influence the probability of mating between susceptible and resistant ECB moths and as such may have a significant influence on the efficiency of the HDR strategy. From that perspective, the present study generated two important findings. First, there was no evidence that the mating success of immigrant, and presumably older, males is any different from that of resident males. Indeed, at the local scale, resident females mated regardless of males having experienced a dispersal event beforehand or not. Limited evidence suggests that, at the study sites, the mating success of immigrant females may, however, be lower than that of resident ones. Second, a fraction of both male and female ECB adults mate before they engage in any long-range dispersal.

Indeed, Experiment 1 shows that virgin females mate indiscriminately with both immigrant and resident males. Our

Bayesian estimate of the proportion of assortative mating among released moths was very low, less than 4%, and its 95% credibility interval included 0, suggesting that, at the local scale, resident females mate randomly with resident and immigrant males. We must, however, caution that this may not be true for resident males, as wild immigrant females carried a significantly lower proportion of marked/total spermatophores than did released females; this could be due to factors such as older age, previous matings, or decreased energy reserves of immigrant females. The generality of this result needs to be confirmed, as this feature was examined over four sessions but for only one site. However, it is consistent with the fact that, conversely, part of the released females mated before dispersal (see below), which means that they, in turn, arrived as mated immigrants in the places to which they dispersed.

Experiment 1 also shows that on average, 4%, but up to 27%, of virgin male and female ECB moths released in a cornfield can be recaptured in the vicinity of this field after two nights, with no evidence for any sex-related difference in recapture rates. Such a delay is known to be long enough for more than 70% ECB females to be mated [22, 42]; and indeed approximately 97% of the females that we recaptured were mated. The proportion of the recaptured males that had mated could not be determined; at least some of them did, as marked spermatophores were recovered from both marked and unmarked females. One of the limitations of this first experiment, though, is that individuals were kept in the laboratory for up to 24 h after emergence before release. Hence, we could not exclude that substantial premating dispersal in fact occurs during a short time window situated very early in adult life but that our experimental design had prevented us from detecting it.

This point was addressed in Experiment 2 using pupae placed directly into the field. The delay between emergence and first flight proved to be relatively long—on average approximately 3.5 h and 9.5 h for moths emerged in the morning and in the evening, respectively—and time intervals between subsequent movements averaged approximately 26 min. The estimated distance covered during this first flight was similar to subsequent flights and was short, on average, less than a meter. Using these values, we calculated that, at dusk, young (less than 24 h old) moths have a 95% probability of being located less than 67 m from their emergence point. Therefore,

Experiment 2 showed no evidence of any high precopulatory dispersal during the very first hours of adult life.

Experiment 2 further suggested that the 4% of moths that were recaptured in Experiment 1 was an underestimate, rather than an overestimate, of the proportion of adult moths remaining locally long enough to mate. This was confirmed by Experiment 3, raising this estimate to an average of approximately 18% for female moths (approximately 34% and 12%, in June and August, respectively). Our experiments thus show that a substantial proportion of recently emerged ECB females can and do mate at a very local scale before engaging in any long-range dispersal. This conclusion is further supported by the fact that 96.2% of wild females captured during the course of Experiment 1 had mated one or several times with wild males, while only 3.8% had mated with released males. The latter estimate is significantly less than the corresponding proportion among released females (19.5%), despite the fact that this proportion was estimated over all sessions for released females but, conservatively, only over the sessions when the proportion of released over total males was highest for wild females. This suggests that a significant part of the wild females captured during the course of this experiment may have preferentially mated at the vicinity of fields from which they emerged and thus before long-range dispersal.

Hence, although the ECB is able to move over far more than a few hundred meters at the adult stage [22–24], its life history may be similar to that of, e.g., Diatraea grandiosella Dyar (Lepidoptera: Crambidae), another stalk-boring moth targeted by Bt corn. In the latter species, more than 91% of females are mated within 24 or 48 h of emergence, 66% of females mate the first night after emergence, and precopulatory flight of males occurs mostly within the natal field (references in [13]). The timing between dispersal and mating has also been studied on another stem borer, Chilo suppressalis, a pest targeted by Bt rice [17]. Approximately 15% of females called and mated within 3 m of the site of eclosion, and approximately 5% of males mated within 5 m of the site of eclosion. In the same vein, Alyokhin and Ferro [16] found that a significant proportion (approximately 50%) of the Colorado potato beetles (Leptinotarsa decemlineata, which is one of the pests targeted by Bt potato) stayed and were seen pairing close to the place of their larval development before they left or before the end of the experiment, but, unlike in the present study, it was not ascertained that the observed pairing behaviour actually resulted in sperm transfer.

What are the implications of the present findings on resistance management? One implicit—and, to our knowledge, so far untested—assumption of most models on which the HDR strategy is based is that, once present locally, immigrant individuals have the same probability of mating with any locally present individual of the other sex, regardless of its having performed dispersal beforehand or not. One of our two main conclusions is that this assumption is valid for immigrant males but questionable for immigrant females. One must keep in mind that our study was performed on susceptible moths. Bt resistance alleles might be associated with fitness costs that may decrease mating success. This has been shown in another lepidopteran pest targeted by Bt crops, Pectinophora gossypiella, where susceptible males were found to mate more often than resistant males in competition for matings with virgin females [43]. The possible impact of Bt resistance on ECB life history traits can currently not be evaluated since, to our knowledge, no ECB strain resistant enough to complete a whole lifecycle on Bt corn has been selected so far.

Our second main conclusion is that some predispersal mating does occur for both males and females. Hence, the current HDR management strategy aimed at delaying resistance to transgenic Bt corn in ECB populations may not ensure complete mixing between susceptible and resistant moths prior to their first mating. This is all the more important as ECB males are thought to mate on average only 3.8 times over their whole lifetime, producing spermatophores of exponentially decreasing size so that the first copulation accounts for approximately 60% of the total volume emitted over their entire lifetime [44].

This lack of complete mixing between susceptible and resistant moths prior to their first mating probably increases the proportion of crossings among homozygous resistant (rr) individuals to be expected. Actually, our results point out that the proportion of rr × rr crossings over the entire landscape strongly depends on the timing and importance of males' long-range dispersal. Indeed, the higher the proportion of SS and rS males emigrating from the refuges into the pool of males present at the edges of Bt cornfields, the lower is the proportion of resident rr females that will be mated by rr males emerged from the same field. Also, it must be noted that the densities we released mimic a field where the frequency of the resistance allele is already quite high. If the r allele is initially as rare as estimated by Bourguet et al. [45], it is possible that resistant ECB in Bt field borders would be at such low densities that their behavior might be different; they might, for instance, leave more readily than in the present study. On the other hand, it is unlikely that an rr individual would be completely alone in a Bt field, as some of its sibs should be present, too. Notably, Experiment 1 shows that, within the pool of males present in the field border, the proportion of males that had immigrated from cornfields located several hundred meters away was very variable between sites and between dates: it ranged from 32.8% to 100%. The spatiotemporal variations of the proportion of resident moths staying in their natal field long enough to mate, and of the local proportion of resident to immigrant males in fields with small resident populations, as well as the factors affecting them appear to be crucial points, that must be studied in more detail. Again, it must also be kept in mind that while these factors need to be studied for susceptible moths, results may not be directly applicable to resistant genotypes. For instance, resistant P. gossypiella have been found to display a longer developmental time [46], which could affect (positively or negatively) the probability of effective mixing at the landscape level.

The influence of predispersal mating on Bt resistance evolution under the HDR strategy is not trivial. While models of the HDR strategy assuming crop rotation conclude that predispersal mating uniformly increases the speed at which Bt resistance alleles are selected [4], others [13, 34, 37, 41] show that, if crops are not rotated, departures from the original assumptions on the life history of the pest (amount of dispersal, timing between dispersal and mating, oviposition, etc.) can also significantly affect the predictions. Notably, they show that the negative consequences of predispersal mating can be offset, provided it is not too high, by non-random oviposition on Bt rather than on non- Bt crop fields. Provided crops are not rotated (i.e., provided the same fields are used as refuges year after year), such preferential oviposition could result from some females staying in their natal field not only to mate but also to oviposit. However, all these models assume, in addition, sex-biased premating dispersal, with females staying in the vicinity of their natal field until mating and males dispersing and then mating randomly with any of them. Our data provide no evidence that such a sex-bias exists, so that, to be conservative, this assumption warrants more careful checking before it can be made.

Some models assuming no crop rotation suggest that an intermediate amount of dispersal can, in certain cases, be better than low or high dispersal: if the HDR strategy were applied in this framework, a precise quantification of the amount of predispersal mating would become important. Our results do not provide such precise quantification. They offer a method to estimate this proportion, but most of all they show that it can be highly variable, suggesting that it may not be so easy to find a single, generally applicable value and that overestimating or underestimating it with a security margin may not be safely applicable either. Therefore, relying on an intermediate level of predispersal mating in the HDR strategy seems to require more finetuning than can probably be concretely achieved on a large scale. In addition, models using this framework assume that female postmating dispersal is low (i.e., that females staying close to their natal field to mate will also stay there to oviposit), an assumption that may be reasonable but remains yet to be checked. Another method for reducing random ovipositing could be strategic cutting of field borders or other means to bias the attractivity of Bt fields versus refuges for oviposition—a "trap-crop" type of strategy already suggested by Alstad and Andow [4].

In sum, our results do not necessarily imply that the HDR strategy cannot work. However, they caution that model assumptions must be carefully investigated before relying on them, and they offer a method to do so, which could also be applied to other pests targeted by Bt crops.

MATERIALS AND METHODS

Statistical analyses were performed using SYSTAT 9.0 software [47] or JMP IN 3.0 software [48]. Throughout this paper, all *p*-values are given for two-tailed tests. Time is given as local time.

Experiment 1.

Mark-release-recapture experiments were conducted during the June 2004 breeding season at four sites located in a corn-growing area, approximately 20 km south of Toulouse, France. Adult ECBs were obtained from an outbred strain that was mass-reared on a wheat diet. Male and female pupae were kept separately, so that adults were all virgin at the time of release, which was conducted less than 24 h after emergence. As wheat and corn use different types of photosynthesis (C3 and C4 type, respectively), the ratio of12C and13C carbon isotopes in their tissues differ [28]. This characteristic is transferred to moth tissues [27], so that laboratory-reared individuals fed on wheat diet can be distinguished from field individuals fed on corn. In addition to this chemical marking, moths were color-marked prior to release. After approximately 30 min at 6 °C (until they were unable to fly), they were marked on the dorsal thorax and base of the wings with a 1:1 ink/ethanol mixture applied with the tip of a matchstick. A different colour was used for each release and for each of the two age classes released (less than 12 h and 12 to 24 h). After marking, moths were placed into small boxes and stored in cool boxes with ice blocks to reduce agitation, which might cause damage or abnormally high dispersal at the time of release. In each site, the experiment was carried out in a field planted with corn in 2004 and with soybean or wheat in 2003. Because these two crops suffer very little infestation by the ECB [26, 27], our experimental set-up mimicked the situation in a cornfield planted with Bt corn the year before.

Herbaceous borders along cornfields are described as a very suitable habitat for the ECB, where large numbers of adults typically mate and rest during the daytime [49]. During the afternoon before each release, a 100-m section of a herbaceous border running along a cornfield and of the first eight rows into the cornfield was cleared of any moth using sweep net capture. Capture efforts were stopped when the 100-m section could be screened once entirely without finding more than five additional moths. At approximately 08:00 P.M., the small boxes containing the marked moths were taken out of the cool boxes, placed open into a 40-m section along a line parallel to the cleared border section, at equal distance from both ends, and situated between the seventh and eighth row into the cornfield, approximately 7 m from the herbaceous border of the field, and gently agitated until the moths left. Recaptures were conducted by sweep net capture, 36 h later, in the 100-m section cleared before release. Again, all efforts were made to ensure exhaustive capture of all marked and unmarked ECB moths over this section. The proportion of recaptured moths was not influenced by their age (less than 12 h or 12 to 24 h) upon release (paired t-test, $t = 0.002$, $df = 15$, $p = 0.998$), so that released moths were pooled regardless of age (less than 24 h) in subsequent analyses.

We dissected all color-marked recaptured females to determine their mating status. We also dissected a subsample (n = 79) of the wild females caught during recapture sessions. In order to estimate their mating rate with marked versus unmarked males conservatively, we took these 79 females from those sessions displaying the highest released/wild males ratios (site 1: 03 June; site 2: 02 June; site 3: 01, 07 and 11 June; site 4: 31 May and 04 June). Finally, for each of four recapture sessions conducted in site 1, we dissected not only the marked but also the unmarked females, and we captured 49 to 55 (unmarked) females in site 5, i.e., in the border of a large cornfield located approximately 400 m away from site 1 that was a likely source of many of the immigrant moths captured there (Tables S3 and S4). The bursa copulatrix of the genital duct of mated female moths contains one or several spermatophores, i.e., solidified droplets of sperm and nutritious substances deposited by males during the mating process (usually one per mating event [44]) and later used by the female. The overall proportion of females mated with wild vs. released males was compared with a Fisher exact test between unmarked and marked females. The numbers of spermatophores carried by marked versus unmarked females at site 1 and by unmarked females at site 1 versus site 5 were compared for each replicate by means of a Kruskal-Wallis nonparametric analysis of variance. The shape and color of spermatophores were used to assign them to two age categories ("recent": white and of ovoid shape, i.e., little digested, versus "old": brown and with a substantial part missing, indicating a more advanced stage of digestion), from which it can be inferred how recently the mating occurred [42, 50]. The proportion of old versus new spermatophores carried by females was compared for each replicate with a Fisher exact test between unmarked and marked females at site 1 and between unmarked females at site 1 and site 5. Global tests across replicates were constructed using Fisher's combined probability test method (Fisher's global test [51, 52]) to compare the number of spermatophores and the proportion of old vs. new spermatophores carried by the same pairs of groups of females.

Stable carbon isotope analysis of wings, legs, or spermatophores of ECB moths were conducted as described in [27]. Results are expressed in conventional "per mill" units relative to the Pee Dee Belemnite standard as: $\delta 13C$ (‰) = [(Rsample − Rstandard)/ Rstandard] × 1,000, where Rsample and Rstandard are the13C/12C atom ratios in the sample and the standard, respectively.

We checked for possible differences in survival and in mating propensity between marked and wild moths in two control experiments: SURVIVAL and MATING. The SURVIVAL experiment was conducted as follows. While we performed Experiment 1, in June 2004, three replicates of five color-marked and five non–color-marked females and five color-marked and five non–color-marked males, all less-than-12-h-old adults from the wheat-fed laboratory strain used in Experiment 1, were placed into a 20 × 20 × 40-cm plastic box with a wet paper wad at room temperature with natural daylight. Deaths were recorded daily. We analysed the number of days of survival using a general linear model with sex, marking, and marking color nested in marking as fixed factors and replicate as a random factor. In laboratory conditions, once the approximately 2-d difference in survival between males and females had been accounted for ($F1,166 = 43.59$, $p < 0.0001$), there was only a slight difference in survival of colour-marked versus unmarked moths, and the difference was in favor of the former (8.32 ± 0.27 d and 7.62 ± 0.26 d, respectively; $F1,166 = 4.67$, $p = 0.032$), and all moths survived at least 2 d (Table S1). Thus, the released moths that were not recaptured in the field were probably mostly still alive.

The MATING experiment was conducted as follows. While we performed Experiment 1, in June 2004, four replicates of ten females and ten males, all color-marked and all from the wheat-fed laboratory strain used in Experiment 1, were placed together with ten unmarked wild males collected less than 2 h earlier, next to a cornfield approximately 5 km from the closest study site, into the same boxes and in the same room conditions as replicates of the SURVIVALexperiment. We thereby controlled that, in laboratory conditions, (i) when caged with equal proportions of marked and unmarked males, marked (this experiment) and non–color-marked (SURVIVAL experiment) females showed the same overall mating rate after 48 h ($\chi 2 = 3.80$, df= 3, $p = 0.284$, Table S1), and that (ii) recently emerged (less than 24 h old) marked males had the same probability to mate with marked females as wild males collected directly in the field (Fisher›s exact test, $p = 0.683$, Table S2). Thus, released females probably had the same propensity to mate as wild females, and so had released and wild males.

Data analysis of assortative mating— we estimated the probability, for a released moth, to mate assortatively with other released moths by assuming that each released virgin female would mate either with a marked, released male or with a wild male present at the study site. Let pA ij and fij be, respectively, the proportion of released females mating assortatively with released males rather than randomly and the proportion of released males among all males present at site i during a given release-and-recapture session j. As an estimate for fij , we used the proportion of marked males among total males captured during the recapture conducted 36 h after the corresponding release). Any female released at site i during session j has a probability pA ij + (1 − pA ij) • fij of mating with a released male and (1 − pA ij)(1 − fij) of mating with a wild male. Using the values of parameters pA ij and fij , it is then possible to compute the likelihood of the observed proportions of marked and unmarked spermatophores retrieved from females captured at each site and for each session. To obtain posterior distributions for pA ij , we applied a Bayesian Markov chain Monte Carlo approach with an uninformative [Uniform (0,1)] prior and lognormal deviates. Ten thousand values were sampled with a thinning of 100 and a burn-in of 10,000 and used to calculate statistics (mode and credibility interval) of the posterior distribution. The 95% credibility interval of pA ij was estimated as the highest posterior density region, i.e., the region of values that contains 95% of the posterior probability [53]. The probability of assortative mating was assumed to be identical for all release sessions and sites.

Experiment 2.

Experiment 2 was conducted during the June 2005 breeding season, in the same corn-growing area as Experiment 1. Approximately 800 pupae were obtained by mass-rearing the same outbred strain used in Experiment 1. These pupae were placed between the seventh and eighth rows of a cornfield during 11 d (05–15 June), i.e., they were allowed to emerge directly in the field. Experiment 2 proceeded in four independent steps, as follows. Over 5 d of observation, we recorded (i) individual emergence times (n = 221), (ii) the time (n = 169), and (iii) the distance (n = 124) of the first flight performed after emergence. After their first flight, moths were gently placed individually in a small plastic box with a wad of wet paper, which was left in field conditions in the shade. We also recorded (iv) the frequency and distance of subsequent flights during a total of 136 observation sessions of 10 min each, spread from 10:00 A.M. to 09:30 P.M. and over 4 d. During these observation sessions, less-than-24-h-old moths that had already performed their first flight were taken out of the small boxes in which they were kept in field conditions and settled individually on cornstalks. Possible moth movement was monitored for 10 min and the distance covered during each flight (if any) was measured and recorded. For data analyses, plots of residuals versus predicted values were examined and the distribution of residuals was checked for normality. To improve normality of the residuals› distribution, the distances covered during the first flight following emergence were Ln (1 + distance) transformed before testing for any differences related to sex and period of emergence.

Experiment 3.

Experiment 3 was conducted, during the first (June) and second (August) 2005 breeding seasons, in the same cornfield as Experiment 2. Females released in June 2005 were obtained from the laboratory strain used in Experiment 2. These females were color-marked as described in Experiment 1. Females released in August 2005 were offspring of approximately 80 wild mated females collected in June 2005 in the vicinity of the experimental site. These offspring were not color-marked but were reared on a wheat diet, which provided a biogeochemical marking allowing later checking.

Females to be released were either kept separate from males since pupal stage or checked very frequently and separated just after emergence, which ensured their not being mated prior to release. Less than 24 h after emergence, they were settled individually on every cornstalk or every second stalk of the eighth row running parallel to the nearest field border. The stalks where released moths were settled were clearly identified with plastic labels. Individual boxes in which moths were kept upon release were gently opened, and the tip of a corn leaf was used to invite the moth to climb the stalk and settle on the plant without flying away. When females flew away, we attempted to recapture them immediately and tried again to settle them until they were either successfully settled or lost. A total of 198 and 549 females were settled successfully in June and August respectively, of approximately 220 and 770 attempts. All females were settled just before sunset—i.e., between approximately 9:00 and 9:30 P.M. in June and between approximately 8:00 and 9:00 P.M. in August, a time when their activity in the field, as well as that of wild females in the vicinity, was usually low (AD, SP, and DB, personal observations). ECBs typically start flying at dusk in search of a suitable site to emit pheromones (females) or of a receptive female (males) and settle at various times before dawn, depending notably on temperature. In the present experiment, moths were checked every 30 to 60 min after release, either until dawn (August) or until the temperature dropped to the point that no moth activity was observed anymore in and around the field for more than 1 h, i.e., approximately 4 h after release (June). Any mating pair including a marked female was collected, and its place of collection (either on cornstalks within the release section, or nearby, within approximately 50 m) was recorded. We checked that all females were able to fly the day after the experiment and that they all survived 3 d or longer in the laboratory. In addition, we confirmed their mating status either by checking for the presence of a spermatophore or by recording the laying of fertile eggs.

ACKNOWLEDGMENTS

We thank D. Fortini (INRA, Le Magneraud) for intensive ECB rearing; S. Meusnier for assistance in the laboratory; T. Malausa, B. Pélissié, and L. Pélozuelo for help during the field experiments; the Bolatti, Martino, Pavant, and Pérès families for permission to work in their cornfields; J.-M. Cornuet for estimating the level of assortative mating; L. Chikhi for helpful discussions; and R. Streiff and three anonymous referees for very helpful comments on the manuscript.

AUTHOR CONTRIBUTIONS

SP, RIB, and DB conceived, designed, and performed Experiment 1. AD, SP, and DB conceived, designed, and performed Experiments 2 and 3. CP substantially contributed to field work in all three experiments. AD, SP, RIB, and DB analysed the data and wrote the paper.

REFERENCES

1. The Clash (1982) Should I stay or should I go? Combat Rock. Epic Rock.
2. Johst K, Brandl R (1997) Evolution of dispersal: The importance of the temporal order of reproduction and dispersal. Proc R Soc Lond Ser B 264: 23–30.
3. Hufbauer RA, Roderick GK (2005) Microevolution in biological control: mechanisms, patterns, and processes. Biol Control 35: 227–239.
4. Alstad DN, Andow DA (1995) Managing the evolution of insect resistance to transgenic plants. Science 268: 1894–1896.
5. Bates SL, Zhao JZ, Roush RT, Shelton AM (2005) Insect resistance management in GM crops: Past, present and future. Nat Biotechnol 23: 57–62.
6. Lenormand T (2002) Gene flow and the limits to natural selection. Trends Ecol Evol 17: 183–189.
7. Bourguet D, Desquilbet M, Lemarié S (2005) Regulating insect resistance management: The case of non- *Bt* corn refuges in the US. J Environ Management 76: 210–220.
8. Zhou X, Kaktor O, Applebaum SW, Coll M (2002) Population structure of the pestiferous moth *Helicoverpa armigera* using RAPD analysis. Heredity 85: 251–256.
9. Han Q, Caprio MA (2004) Restricted gene flow among local populations of *Heliothis virescens* (F.) during the cotton growing season: Evidence from allozyme and RAPD markers. Environ Entomol 33: 1223–1231.
10. Kim KS, Sappington TW (2005) Genetic structuring of Western corn rootworm (Coleoptera: Chrysomelidae) populations in the U.S. based on microsatellite loci analysis. Env Entomol 34: 494–503.
11. Rousset F (2003) Spatially homogeneous dispersal: the island model and isolation by distance. Genetic structure and selection in subdivided

populations. Princeton (New Jersey): Princeton University Press. pp. 23–52.

12. Ferré J, Van Rie J (2002) Biochemistry and genetics of insect resistance to *Bacillus thuringiensis*. Annu Rev Entomol 47: 501–533.
13. Caprio MA (2001) Source-sink dynamics between transgenic and non-transgenic habitats and their role in the evolution of resistance. J Econ Entomol 94: 698–705.
14. Schneider JC, Roush RT, Kitten WF, Laster ML (1989) Movement of *Heliothis virescens*(F.) (Lepidoptera: Noctuidae) in Mississippi in the spring: Implications for area-wide management. Environ Entomol 18: 438–446.
15. King ABS, Armes NJ, Pedgley DE (1990) A mark-capture study of *Helicoverpa armigera* dispersal from pigeonpea in southern India. Entomol Exp Appl 55: 257–266.
16. Alyokhin AV, Ferro DN (1999) Reproduction and dispersal of summer-generation Colorado poptato beetle (Coleoptera: Chrysomelidae). Environ Entomol 28: 425–430.
17. Cuong NL, Cohen MB (2003) Mating and dispersal behaviour of *Scirpophaga incertulas*and *Chilo suppressalis* (Lepidoptera; Pyralidae) in relation to resistance management for rice transformed with *Bacillus thuringiensis* toxin genes. Int J Pest Management 49: 275–279.
18. Cameron PJ, Wallace AR, Madhusudhan VV, Wigley PJ, Qureshi MS, et al. (2005) Mating frequency in dispersing potato tuber moth, *Phthorimaea operculella,* and its influence on the design of refugia to manage resistance in Bt transgenic crops. Environ Entomol 115: 323–332.
19. Gianessi LP, Carpenter JE (1999) Agricultural biotechnology: Insect control benefits. National Center for Food Agricultural Policy Reports. Available:http://www.ncfap.org/reports/biotech/insectcontrolbenefits.pdf. Accessed 24 Nov 2005 .
20. US Environmental Protection Agency (2001) Biopesticides registration action document: Bacillus thuringiensis plant-incorporated protectants. Available:http://www.epa.gov/pesticides/biopesticides/pips/bt_brad.htm. Accessed 24 Nov 2005 .
21. Showers WB, Reed GL, Robinson JF, Derozari MB (1976) Flight and sexual activity of the European corn borer. Environ Entomol 5: 1099–1104.
22. Hunt TE, Higley LG, Witkowski JF, Young LJ, Hellmich RL (2001) Dispersal of adult European corn borer (Lepidoptera: Crambidae) within and proximal to irrigated and non-irrigated corn. J Econ

Entomol 94: 1369–1377.

23. Showers WB, Hellmich RL, Derrick-Robinson ME, Hendrix WH (2001) Aggregation and dispersal behavior of marked and released European corn borer (Lepidoptera: Crambidae) adults. Environ Entomol 30: 700–710.
24. Qureshi JA, Buschman LL, Throne JE, Ramaswamy SB (2005) Adult dispersal of*Ostrinia nubilalis* Hübner (Lepidoptera: Crambidae) and its implications for resistance management in *Bt*-maize. J Appl Entomol 129: 281–292.
25. Rubenstein DR, Hobson KA (2004) From birds to butterflies: Animal movement patterns and stable isotopes. Trends Ecol Evol 19: 256–263.
26. Caffrey D, Worthley L (1927) A progress report on the investigations of the European corn borer. U S Dept Agri Bull 1476: 1–155.
27. Ponsard S, Bethenod MT, Bontemps A, Pélozuelo L, Souqual MC, et al. (2004) Carbon stable isotopes: A tool for studying the mating, oviposition, and spatial distribution of races of European corn borer, *Ostrinia nubilalis*, among host plants in the field. Can J Zool 82: 1177–1185.
28. DeNiro MJ, Epstein S (1978) Influence of diet on the distribution of carbon isotopes in animals. Geochim Cosmochim Ac 42: 495–506.
29. Gould F, Blair N, Reid M, Rennie TL, Lopez J, et al. (2002) *Bacillus thuringiensis*toxin resistance management: Stable isotope assessment of alternate host use by*Helicoverpa zea*. Proc Natl Acad Sci U S A 99: 16851–16586.
30. Bontemps A, Bourguet D, Pélozuelo L, Bethenod M-T, Ponsard S (2004) Managing the evolution of *Bacillus thuringiensis* resistance in natural populations of the European corn borer, *Ostrinia nubilalis*: Host plant, host race and pherotype of adult males at aggregation sites. Proc R Soc Lond Ser B 271: 2179–2185.
31. Malausa T, Bethenod M-T, Bontemps A, Bourguet D, Cornuet JM, et al. (2005) Assortative mating in sympatric host races of the European corn borer. Science 208: 358–260.
32. Royer L, McNeil JN (1991) Changes in calling behaviour and mating success in the European corn borer (*Ostrinia nubilalis*), caused by relative humidity. Entomol Exp Appl 61: 131–138.
33. Gould F (1998) Sustainability of transgenic insecticidal cultivars: Integrating pest genetics and ecology. Annu Rev Entomol 43: 701–726.
34. Peck SL, Gould F, Ellner SP (1999) Spread of resistance in spatially extended regions of transgenic cotton: Implications for management of *Heliothis virescens* (Lepidoptera: Noctuidae). J Econ Entomol 92:

1–16.

35. Carrière Y, Tabashnik BE (2001) Reversing insect adaptation to transgenic insecticidal plants. Proc R Soc Lond Ser B 268: 1475–1481.
36. Guse CA, Onstad DW, Buschman LL, Porter P, Higgins RA, et al. (2002) Modeling the development of resistance by stalk-boring Lepidoptera (Crambidae) in areas with irrigated transgenic corn. Environ Entomol 31: 676–685.
37. Ives AR, Andow DA (2002) Evolution of resistance to *Bt* crops: Directional selection in structured environments. Ecol Lett 5: 792–801.
38. Onstad DW, Guse CA, Porter P, Buschman LL, Higgins RA, et al. (2002) Modeling the development of resistance by stalk-boring lepidopteran insects (Crambidae) in areas with transgenic corn and frequent insecticide use. J Econ Entomol 95: 1033–1043.
39. Vacher C, Bourguet D, Rousset F, Chevillon C, Hochberg ME (2003) Modelling the spatial configuration of refuges for a sustainable control of pests: A case study of *Bt*cotton. J Evol Biol 16: 1–10.
40. Vacher C, Bourguet D, Rousset F, Chevillon C, Hochberg ME (2004) High dose refuge strategies and genetically modified crops - reply to Tabashnik et al. J Evol Biol 17: 913–918.
41. Heimpel GE, Nauhauser C, Andow DA (2005) Natural enemies and the evolution of resistance to transgenic insecticidal crops by pest insects: The role of egg mortality. Environ Entomol 34: 512–526.
42. Showers WB, Reed GL, Oloumi-Sadeghi H (1974) Mating studies of female European corn borers: Relationship between deposition of egg masses on corn and captures in light traps. J Econ Entomol 67: 616–619.
43. Higginson DM, Morin S, Nyboer ME, Biggs RW, Tabashnik BE, et al. (2005) Evolutionary trade-offs of insect resistance to *Bacillus thuringiensis* crops: Fitness cost affecting paternity. Evolution 59: 915–920.
44. Royer L, McNeil JN (1993) Male investment in the European corn borer, *Ostrinia nubilalis* (Lepidoptera: Pyralidae): Impact on female longevity and reproductive performance. Funct Ecol 7: 209–215.
45. Bourguet D, Chaufaux J, Séguin M, Buisson C, Hinton JL, et al. (2003) Frequency of alleles conferring resistance to *Bt* maize in French and US corn belt populations of the European corn borer, *Ostrinia nubilalis*. Theor Appl Genet 106: 1225–1233.
46. Liu TX (1999) Effects of some new formulations of *Bacillus thuringiensis* for management of cabbage looper and diamondback moth on cabbage in South Texas. Southwestern Entomol 24: 167–177.

47. SPSS Inc (1998) SYSTAT 9.0 ed. statistics. Chicago: SPSS.

48. SAS Institute (1996) JMP IN 3.0 ed. Cary (North Carolina): SAS Institute.

49. Pleasants JM, Bitzer RJ (1999) Aggregation sites for adult European corn borers (Lepidoptera: Crambidae): A comparison of prairie and non-native vegetation. Environ Entomol 28: 608–617.

50. Elliott WM, Dirks VA (1979) Postmating age estimates for female European corn borer moths, *Ostrinia nubilalis* (Lepidoptera: Pyralidae), using time-related changes in spermatophores. Can Entomol 111: 1325–1335.

51. Fisher R (1932) Statistical methods for research workers. Edinburgh: Oliver and Boyd.

52. Whitlock MC (2005) Combining probability from independent tests: The weighted Z-method is superior to Fisher's approach. J Evol Biol 18: 1368–1373.

53. Gelman A, Carlin JB, Stern HS, Rubin DB (1995) Bayesian data analysis. Norwell (Massachusetts): Chapman & Hall. 526 p.

54. Bourguet D, Bethenod MT, Trouvé C, Viard F (2000) Host-plant diversity of the European corn borer *Ostrinia nubilalis*: What value for sustainable transgenic insecticidal *Bt* maize? Proc R Soc Lond Ser B 267: 1177–1184.

55. Pélozuelo L, Malosse C, Genestier G, Guenego H, Frérot B (2004) Host-plant specialization in pheromone strains of the European corn borer *Ostrinia nubilalis* in France. J Chem Ecol 30: 335–351.

56. Bethenod MT, Thomas Y, Rousset F, Frérot B, Pélozuelo L, et al. (2005) Genetic isolation between two sympatric host plant races of the European corn borer, *Ostrinia nubilalis* Hübner. II. Assortative mating and host plant preferences for oviposition. Heredity 94: 264–270.

Chapter 5

COMPLETE CHLOROPLAST GENOME SEQUENCES OF IMPORTANT OILSEED CROP SESAMUM INDICUM L

Dong-Keun Yi, Ki-Joong Kim

School of Life Sciences, Korea University, Seoul, Korea

ABSTRACT

Sesamum indicum is an important crop plant species for yielding oil. The complete chloroplast (cp) genome of *S. indicum* (GenBank acc no. JN637766) is 153,324 bp in length, and has a pair of inverted repeat (IR) regions consisting of 25,141 bp each. The lengths of the large single copy (LSC) and the small single copy (SSC) regions are 85,170 bp and 17,872 bp, respectively. Comparative cp DNA sequence analyses of *S. indicum* with other cp genomes reveal that the genome structure, gene order, gene and intron contents, AT contents, codon usage, and transcription units are similar to the typical angiosperm cp genomes. Nucleotide diversity of the IR region between *Sesamum* and three other cp genomes is much lower than that of the LSC and SSC regions in both the coding region and noncoding region. As a summary, the regional constraints strongly affect the sequence evolution of the cp genomes, while the functional constraints weakly affect the sequence evolution of cp genomes. Five short inversions associated with short palindromic sequences that form step-loop structures were observed in the chloroplast genome of *S. indicum*. Twenty-eight different simple sequence repeat loci have been detected in the chloroplast genome of *S. indicum*. Almost all of the SSR loci were composed of A or T, so this may also contribute to the A-T richness of the cp genome of *S. indicum*. Seven large repeated loci in the chloroplast genome of *S. indicum* were also identified and these loci are useful to developing *S. indicum*-specific cp genome vectors. The complete cp DNA sequences of *S. indicum* reported in this paper are prerequisite to modifying this important oilseed crop by cp genetic engineering techniques.

INTRODUCTION

Sesamum indicum L. cv. Ansanggae is an annual plant that reaches 50 to 100 cm tall. Commonly known as sesame or til, *S. indicum* is an important and ancient oil-yielding crop. Sesame seeds are cultivated as a rich source of edible oil. Though the origin is uncertain, the species probably originated from southeastern Africa and a naturalized population has been found in India [1], [2]. *S. indicum* is widely cultivated and naturalized in dry habitats of tropical and subtropical regions, with the primary production occurring in the developing countries of Asia and Africa. Sesame seeds contain approximately 50–60% edible oil, and sesame oil is ranked 5th in terms of oil production. Sesame seed production worldwide is estimated at 4 million tons and production is steadily growing. Nearly 70% of the world's production is consumed in the producing countries and world trade is limited. The major producing countries in descending order are: Myanmar (867,520 tons), India (657,000 tons), China (622,905 tons), Sudan (318,000 tons), Ethiopia (260,534 tons), Uganda (178,000 tons), and Nigeria (110,000 tons). *S. indicum* is a member of the family Pedaliaceae, order Lamiales. This order also includes the family Oleaceae.

In comparison with sunflower, canola, and soybeans, crops which are primarily cultivated in advanced countries, the modern genetic research on sesame has been relatively limited [3], [4]. This is because *S. indicum* is mostly cultivated in developing countries. Most studies have focused on the nutrients and products of sesame. Most recently, ISSR markers and EST tags were developed for the creation of genetic maps in sesame [5], [6], [7]. To date, there have been no known studies of the sesame chloroplast (cp) genome sequence. Therefore, the complete cp genome sequences of *S. indicum* were generated and characterized for their suitability as cp genome vector sequences for application in future genetic engineering studies.

The genomes of chloroplasts, the plant organelles responsible for photosynthesis [8], [9], provide rich evolutionary and phylogenetic information [10], [11]. Accordingly, several recent studies have used cp genome information to construct the angiosperm phylogeny [12], [13]. The complete cp genomes of more than 170 species, including many crop species, have been reported from various groups of plants and algae (Chloroplast Genome Database,http://chloroplast.cbio.psu.edu).

The majority of the cp genomes of land plants contain 90–110 unique genes within the 115–165 kb of circular chromosome [14]. The primary mechanism of gene order change is inversion by intramolecular recombination, and this method occurs principally via the dispersed repeats of the cp genome [15], [16]. Evolutionary hot spots showing high levels of insertions and deletions (indels) with high incidences of base substitutions are concentrated on specific gene and intergenic spacers [17]. Several comparative studies have documented the phylogenetic usefulness of cp genome structures at higher taxonomic levels [18], [19]. However, only a few studies have explored the usefulness of cp genome data in closely related taxa.

Currently, transformation using chloroplast vectors provides a valuable technique for chloroplast genetic engineering [20]. Cp genome vectors show high-levels of gene expression, the possibility of the expression of multiple genes or pathways via a single transformation event, and transgene containment due to a lack of pollen transmission [21]. Gene transformation protocols using universal cp genome vectors have been developed in tobacco and carrots [22], [23]. However, the universal vectors show limited utility for distantly related plant species. To construct a species-specific cp genome vector, the complete cp DNA sequence is necessary. For this purpose, the complete cp DNA sequences from *S. indicum* (Pedaliaceae) are reported herein. In addition, a comparative sequence analysis of the whole cp genomes of *S. indicum* and*O. europaea* was conducted to reveal more information concerning recent cp genome evolution. The comparative data will contribute to an increased understanding of the evolutionary model of the cp genome in the order Lamiales. To develop gene transformation protocols using a chloroplast genome vector, an analysis was performed on the repeating sites within the chloroplast genome. This information may enable the production of sesame-specific chloroplast genome vectors.

Results

General Features of the *Sesamum Indicum* cp Genome

The *Sesamum indicum* cp genome exhibits the general cp genome structure characteristic of flowering plants. It contains a pair of inverted repeat regions (IRa and IRb) that comprise 25,141 bp each. The two IR regions divide the genome into a large single copy (LSC) region and a small single copy (SSC) region. The LSC region is 85,170 bp, whereas the SSC region is 17,872 bp. The complete cp sequence of *S. indicum* is 153,324 bp in length (GenBank acc no. JN637766), of which 58% is coding regions and 42% is non-coding regions. A total of 114 genes are contained within the *S. indicum* cp genome, including 80 protein-coding genes, 30 transfer RNA genes, and four ribosomal RNA genes (Figure 1, Table 1). Ten protein-coding and seven tRNA coding genes are duplicated on the IR regions. The LSC region contains 62 protein-coding and 22 tRNA genes, while the SSC region contains 12 protein-coding and one tRNA gene. Similar to the *Nicotiana* and *Panax* cp genomes, 18 of the genes in the *S. indicum* cp genome have one or two introns. Of these, *rps12, clpP* and *ycf3* have two introns. The *rps12*gene is a uniquely divided gene with the 5′ end exon located in the LSC region while two copies of 3′ end exon and intron are located in the IR region.

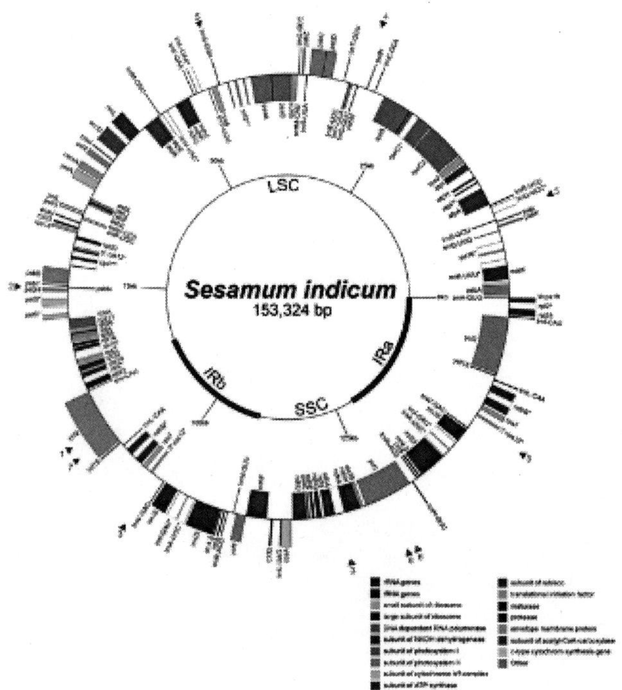

Figure 1. The gene map of Sesamum indicum cp genome. A pair of thick lines at the inside circle represents the inverted repeats (IRa and IRb; 25,141 bp each), which separate the large single copy region (LSC; 85,170 bp) from the small single copy region (SSC; 17,872 bp). Genes drawn inside the circle are transcribed clockwise, while those drawn outside the circle are transcribed counterclockwise. Intron-containing genes are marked by asterisks. The numbers at the outmost circle indicate the locations of 7 repeats including direct (black number), palimdromic (blue number), and dispersed repeats (red numbers), respectively (cf. Table 4).

doi:10.1371/journal.pone.0035872.g001

Table 1. Genes contained in the Sesamum indicum cp genome (total 114 genes).

Category for genes	Group of genes	Name of genes
Self replication	rRNA genes	rrn16(x2), rrn23(x2), rrn4.5(x2), rrn5(x2)
	tRNA genes	30 trn genes (6 contain an intron, 7 in the IR regions)
	Small subunit of ribosome	rps2, rps3, rps4, rps7(x2), rps8, rps11, rps12*, rps14, rps15, rps16*, rps18, rps19
	Large subunit of ribosome	rpl2*(x2), rpl14, rpl16*, rpl20, rpl22, rpl23(x2), rpl32, rpl33, rpl36
	DNA dependent RNA polymerase	rpoA, rpoB, rpoC1*, rpoC2
Genes for photosynthesis	Subunits of NADH-dehydrogenase	ndhA*, ndhB*(x2), ndhC, ndhD, ndhE, ndhF, ndhG, ndhH, ndhI, ndhJ, ndhK
	Subunits of photosystem I	psaA, psaB, psaC, psaI, psaJ, ycf3**
	Subunits of photosystem II	psbA, psbB, psbC, psbD, psbE, psbF, psbH, psbI, psbJ, psbK, psbL, psbM, psbN, psbT, psbZ
	Subunits of cytochrome b/f complex	petA, petB*, petD*, petG, petL, petN
	Subunits of ATP synthase	atpA, atpB, atpE, atpF*, atpH, atpI
	Large subunit of rubisco	rbcL
Other genes	Translational initiation factor	infA
	Maturase	matK
	Protease	clpP**
	Envelope membrane protein	cemA
	Subunit of Acetyl-CoA-carboxylase	accD
	c-type cytochrom synthesis gene	ccsA
Genes of unknown functions	Open Reading Frames (ORF, ycf)	ycf1, ycf2(x2), ycf4, ycf15(x2)

One and two asterisks after gene names reflect one- and two-intron containing genes, respectively. Genes located in the IR regions are indicated by the (x2) symbol after the gene name. The rps12 gene is divided; the 5'-rps12 is located in the LSC region and the 3'-rps12 in the IR region.
doi:10.1371/journal.pone.0035872.t001

doi:10.1371/journal.pone.0035872.t001

The overall GC and AT contents of the *S. indicum* cp genome are 38% and 62%, respectively. The AT content of the IR regions (57%) is lower than that of the LSC and SSC regions (64% and 68%, respectively). This low AT content in the IR regions is due to the low AT content of four rRNA genes in the region: *rrn16*, *rrn23*, *rrn4.5*, and *rrn5*. The AT content of the protein-coding regions is 60%. Within protein coding region, the AT content is 53% for the first codon position, 62% for the second position, and 70% for the third position, respectively (Table 2). The*Sesamum indicum* cp genome contains 30 tRNA genes that interact with 20 amino acids. Six of the 30 tRNA genes (*trn*K-UUU, *trn*G-UCC, *trn*L-UAA, *trn*V-UAC, *trn*I-GAU and *trn*A-UGC) contain an intron within the anticodon step/loop or D-stem regions.

Table 2. Base compositions in the Sesamum indicum cp genome.

		T(U)	C	A	G	Sequence length(bp)
LSC region		32.5%	18.6%	31.7%	17.8%	85,170
IRa region		28.4%	22.5%	28.2%	20.9%	25,141
IRb region		28.2%	20.9%	28.4%	22.5%	25,141
SSC region		33.6%	17.8%	33.8%	15.5%	17,872
Total		31.3%	19.4%	30.5%	18.8%	153,324
Protein coding genes (CDS)		31.5%	17.6%	30.4%	20.5%	68,697
	1st position	23.0%	18.8%	30.2%	27.5%	22,899
	2nd position	33.0%	20.6%	28.8%	18.8%	22,899
	3rd position	38.0%	13.6%	32.7%	16.1%	22,899

doi:10.1371/journal.pone.0035872.t002

doi:10.1371/journal.pone.0035872.t002

The length of angiosperm cp genomes is variable primarily due to expansion and contraction of the inverted repeat IR region and the single copy boundary regions. To elucidate this mechanism, the IR/SC boundary regions of the cp genomes of *Sesamum, Nicotiana, Panax, Olea,* and *Arabidopsis* (Figure 2) were compared. *Rps19* and *ycf1* pseudogenes of various lengths were found at the IR/LSC and IR/SSC boundaries, respectively. The *rps19* pseudogene was not found at the LSC region in the *Nicotiana* and *Olea* cp genomes. In the *Sesamum* cp genome, the IR extended into the *rps19* gene and created a short *rps19* pseudogene of 30 bp at the IR/LSC border. This same pseudogene was 51 bp and 113 bp, respectively, in the *Panax*and *Arabidopsis* cp genomes. At the IR/SSC border of the *Sesamum* cp genome, the IR extended into the *ycf1* gene to create a long *ycf1* pseudogene of 1,100 bp at the IR/LSC border. This *ycf1* pseudogene was 1,164 bp in *Olea* and 1,649 bp in *Panax*. In addition, the *ycf1*pseudogene and the *ndh*F gene overlap in both the *Olea* and *Sesamum* cp genomes for 97 and 70 bp, respectively.

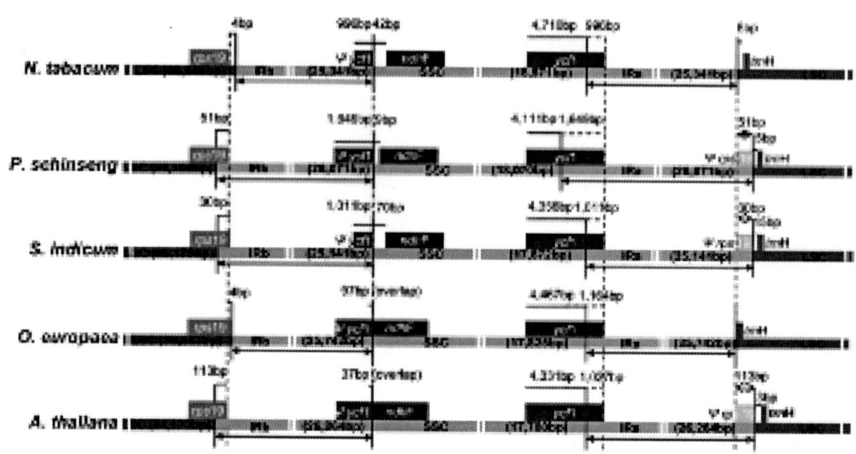

Figure 2. The comparison of the LSC, IR and SSC border regions among five cp genomes.

doi:10.1371/journal.pone.0035872.g002

A comparison of base substitutions and indels in the cp genomes of *Sesamum, Olea, Nicotiana*and *Panax* was conducted. The average sequence divergence of the IR regions is 0.91% between *Sesamum* and *Olea*, 1.52% between *Sesamum* and *Nicotiana*, and 1.67% between*Sesamum* and *Panax*. The divergence values of the LSC regions are 3.23%, 5.64% and 6.17%, respectively, while the divergence values of the SSC regions are 7.06%, 11.25%, 11.98%, respectively. The detailed sequence comparisons in each gene coding region among *Sesamum,Olea, Nicotiana*, and *Panax* are provided in supplemental data (Table S1, S2, and S3). The average Ka/Ks ratios were 0.73, 0.63 and 0.61 in the IR region; 0.23, 0.17 and 0.27 in the LSC region; and 0.40, 0.35 and 0.41 in the SSC region. We also compare the sequence divergence according to the functional groups of genes. The rRNA gene group in the IR region showed the most conserved nature. In contrast, the *matK, ccsA, accD, ycf(5), infA*, and *cemA* genes exhibit high sequence divergences. *Rps*16 and *rpl*33 genes showed a Ka/Ks ratio greater than 1.00 in*Sesamum* and *Olea* comparision.

Sesamum cp genome contain 128 intergenic spacer (IGS) regions which longer than 10 bp in length. The indel and base substitution patterns of the IGS regions were compared among the four cp genomes (Table S1, S2, and S3). The sequence divergences of IGS regions of*Sesamum* and *Olea* ranged from 0.00% to 11.67% in the IR region, 0.00% to 23.18% in the LSC region, and 0.00% to 13.69% in the SSC region, respectively. The divergence values between*Sesamum* and *Panax* ranged from 0.00% to 7.55% in the IR region, 0.00% to 31.52% in the LSC region and 0.00% to 23.91% in the SSC region, respectively. The divergence values between*Sesamum* and *Nicotiana* ranged from 0.00% to 11.11% in the IR region, 0.00% to 50.00% in the LSC region, and 0.00% to 24.55% in the SSC region, respectively. The sequence divergence patterns in the 19 intron regions are provided in Figure 3.

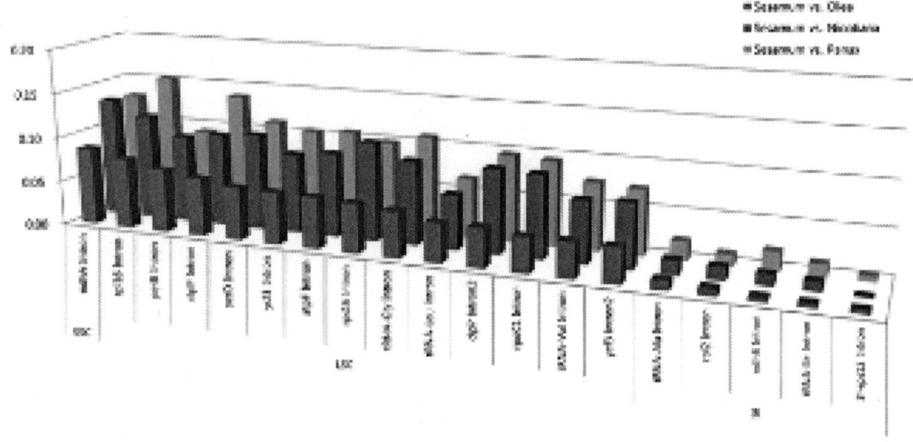

Figure 3. The comparisons of 19 intron regions of the chloroplast genomes in the three different comparisons of Sesamum vs. Olea, Sesamum vs. Nicotiana, and Sesamum vs. Panax. Y axis indicates the sequence divergences.

doi:10.1371/journal.pone.0035872.g003

Upon comparison with the *Olea* cp genome, five short inversions that were associated with inverted sequences were identified in the *Sesamum* cp genome (Figure 4). These five regions form distinct stem-loop hairpin structures, and the sequence orientation is opposite in the two chloroplast genomes at the loop regions. The first short inversion is located on the *rpoB* coding region, genome coordinates 26,401 bp–26,418 bp (Figure 4a). The other four small inversions are located on the inter-genic spacers (Figure 4b–4e). The small inversion regions correspond to the stem-loop-forming regions located downstream of the genes involved in stabilizing mRNA molecules. Large inversion mutations have been frequently noted in several widely diverse vascular plants [24], [25], [26], [27], [28]. In contrast, the short inversions have been recently reported in just a few cp genomes [29], [30], [31], [32].

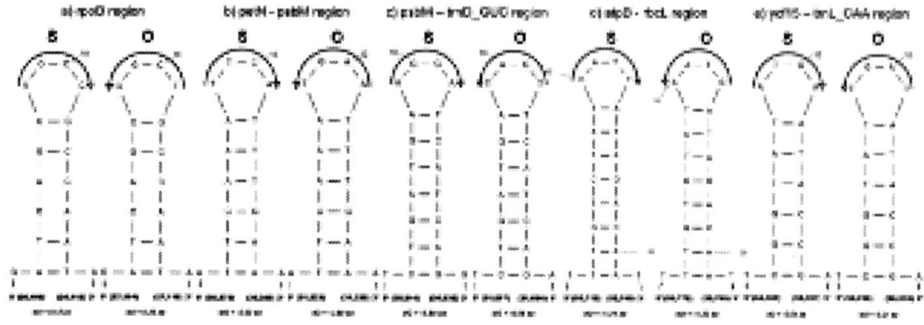

Figure 4. Small inversion mutations and associated secondary structures between the cp genomes of Sesamum (S) and the cp genome of Olea (O).

doi:10.1371/journal.pone.0035872.g004

Simple sequence repeats (SSRs), also called microsatellites, are considered valuable molecular markers for population genetics because they exhibit high variation within the same species[33], [34]. SSRs are stretches of one to six nucleotide units repeated in tandem and randomly spread throughout cp genomes. SSRs are highly polymorphic due to a high mutation rate that affects the number of repeat units. Within the *Sesame* cp genome, 28 different SSR loci are repeated more than 10 times (Table 3). Of these, 21 loci are homopolymers, four are di-polymers, and three are tri-polymers. Eighteen of the homopolymer loci contain multiple A or T nucleotides, while the other three homopolymer loci contain multiples of C or G nucleotides. All of the di-polymer loci contain multiple AT or TA. These SSR loci contribute to the A-T richness of the cp genome of *Sesamum*. Twenty-three SSR loci occur in the intergenic spacers, while only five are located in the gene coding regions of *atpB*, *rpoC2*, *psbC* and *ycf1*.

Table 3. Distribution of simple sequence repeat (SSR) loci in the Sesamum indicum cp genome.

Base	Length	No. SSRs	Coordinated Basepairs[a]
A	13	6	219–248, 4,351–4,399, 8,578–8,587, 72,464–72,475, 121,267–121,278, 133,512–133,521
C	13	2	51,900–51,912, 66,323–66,335
C	11	1	36,699–36,709
T	13	11	**18,418–13,4[illegible]**, **18,[illegible]–18,[illegible](rpoC2)**, 41,[illegible]–41,[illegible], 44,795–44,805, 49,[illegible]–49,[illegible], 55,417–55,42[illegible](rps[illegible]), 58,974–59,983, 71,[illegible]–71,[illegible], 102,[illegible]–102,[illegible], 118,[illegible]–118,[illegible], 126,385–126,40[illegible](ycf1)
T	11	1	61,161–61,171
AT	13	1	20,268–20,277(rpoC2)
AT	12	1	42,984–42,995
TA	13	1	31,905–31,914
TA	12	1	**47,640–47,651**
ATA	12	1	53,471–53,482
TTA	12	1	33,018–33,029
TTC	12	1	35,684–35,695(psbC)

The coordinated basepairs are the nucleotide number positions starting at the IRa/LSC junction (Figure 1). The underline represents the SSR in the CDS and the bold numbers represent the shared SSR with Olea chloroplast genome.
doi:10.1371/journal.pone.0035872.t003

doi:10.1371/journal.pone.0035872.t003

The coordinated basepairs are the nucleotide number positions starting at the IRa/LSC junction (Figure 1). The underline represents the SSR in the CDS and the bold numbers represent the shared SSR with *Olea* chloroplast genome.

Repeats of 26 bp or longer and with sequence identity greater than 90% were also examined. The majority of these were tandem repeats. The repeating unit, repeating time, repeating location, and the total repeating length of the long repeats were evaluated using the Tandem Repeat Finder. From this analysis, seven total repeats were identified and located. This included two direct tandem repeats, two direct inverted repeats, and three palindromic dispersed repeats as possible gene introduction sites (Table 4). The repeating units are repeated two to four times. One dispersed repeat occurs in the widely separated IR and SSC regions of the *Sesamum* cp genome (Table 4).

Table 4. Distribution of large repeat loci in the Sesamum indicum cp genome.

Repeat Number	Size (bp)	Repeat	Location	Repeat Unit	Region
1	72	direct	CDS(ycf2)	[illegible] (4×)	IRb,a
2	49	palindromic	IGS(psbE, psbM)	[illegible]	LSC
3	41	palindromic dispersed repeats	[illegible]	[illegible]	IR,SSC
4	33	palindromic	IGS(petN, psbM)	[illegible]	LSC
5	30	palindromic dispersed repeats	[illegible]	[illegible]	LSC
6	30	palindromic dispersed repeats	CDS(ycf1)	[illegible] (2×)	IRb,a
7	26	direct	CDS(ycf2)	[illegible] (2×)	IRb,a

[illegible]

doi:10.1371/journal.pone.0035872.t004

PHYLOGENETIC ANALYSIS OF SESAMUM BASED ON THE COMPLETE CP GENOME SEQUENCES

In order to identified the phylogenetic position of *Sesamum* within the asterid lineages, 32 complete cp genome sequences are downloaded from the Genbank of NCBI database. Two additional eudicot cp genome sequences from *Spinacia* and *Arabidopsis* also included in the phylogenetic analysis as outgroup taxa. The 24 of 32 complete cp DNA sequences are concentrated in the four major families of asterids such as Solanaceae(7), Oleaceae(6), Apiaceae(6), and Asteraceae(5). Other seven sequences represent Convolvulaceae (*Ipomaea*), Pedaliaceae (*Sesamum*), Rubiaceae (*Coffea*), Araliaceae (*Panax* and *Hydrocotyle*), Goodeniaceae (*Scaevola*) and Campanulaceae (*Trachelium*), respectively. We aligned all protein coding gene sequences and four *rrn* gene sequences in a single data matrix. All *trn*genes are excluded in alignment. The aligned data matrix consists of 83,072 bp in length. About 46% of sites are constant, while the other 54% of sites are variable in sequences or indels.

A maximum likelihood tree was obtained with the likelihood value of -lnL = 428640.9970 with the GTR+G+I base substitution model (Figure 5). The majority of clades are supported by the high levels of Bayesian percentages. We also estimated the splitting times of major clades of asterids using molecular clocks. Two internal fossil data (Araliaceae 70 million years ago (mya) and Solanaceae 53 mya) were used to calibration the clock [35], [36]. The resulting tree indicate that*Sesamum* (Pedaliaceae) form a sister group to the Oleaeaeae (represented by *Jsaminum* and*Olea*) and the two lineages diverged at the Cretaceous-Tertiary (K-T) boundary in geological time (Figure 5).

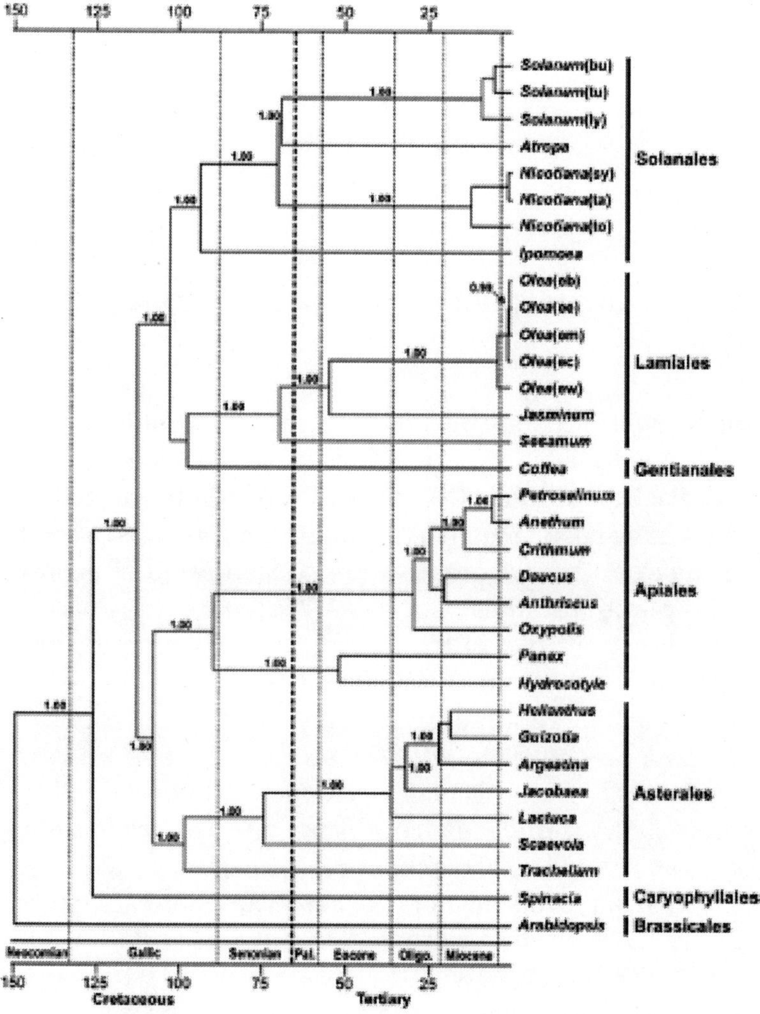

Figure 5. A maximum likelihood tree (-lnL = 428640.9970) of the asterid clade of angiosperm using whole chloroplast genome sequences. The tree was polarized by two outgroup taxa, *Spinacea* and *Arabidopsis*. The GTR+G+I base

substitution model was adopted based on the Modeltest. Molecular clock was calibrated using two internal splitting points of the members of Araliaceae (70 mya) and Solanaceae (53 mya). The numbers above each node indicate the Bayesian support percentages. Taxon abbreviations are Solanum(bu): *Solanum bulbocastanum*,Solanum(tu): *Solanum tuberosum*, Solanum(ly): *Solanum lycopersicum*, Nicotiana(sy):*Nicotiana sylvestris*, Nicotiana(ta): *Nicotiana tabacum*, Nicotiana(to): *Nicotiana tomentosiformis*, Olea(eb): *Olea europaea cv. bianchera*, Olea(ee): *Olea europaea*subsp. *europaea*, Olea(em): *Olea europaea* subsp. *maroccana*, Olea(ec): *Olea europaea* subsp. *Cuspidate* and Olea(ew): *Olea europaea* subsp. *Woodiana*,respectively.

doi:10.1371/journal.pone.0035872.g005

DISCUSSION

Comparative Analysis of the cp Genomes' Structure and Gene Order

Sesamum indicum is an important oilseed crop that is cultivated worldwide for its high quality edible oil. Approximately 170 completed cp DNA sequences have been reported (NCBI GenBank). Of these, 100 complete cp genomes have been sequenced from various groups of seed plants; however, most of these sequences are concentrated in economically important plant families such as Solanaceae, Poaceae, and Asteraceae. For example, of the 24 complete cp genomes published in Asterids, nine are from the Solanaceae family. In contrast, only three complete cp genome sequences have been reported from Lamiales, and no complete cpDNA sequences have been reported for Lamiaceae *s.l.*, which includes *Sesamum*. The availability of the complete cp DNA sequences from *Sesamum* provides us an improved evolutionary understanding of the chloroplast genome itself and it also serves as an agronomic improvement tool. The complete chloroplast genome of *S. indicum* is 153,324 bp long with an LSC region of 85,170 bp, a SSC region of 17,872 bp, and two IR regions of 25,141 bp each (Figure 1). Overall, the genome order, the genome size, the gene and intron contents, and the AT compositions of the *Sesamum* cp DNA show the characteristics typical of land plant cp genomes (Tables 1–2); however, the IR expansion/contraction in the *Sesamum* cp genome generates slightly different pseudogenes in the boundaries (Figure 2). This is not unusual as slight IR expansion/contraction is relatively common in other cp genomes [37], [38].

Analysis of Evolutionary Constraints in the *Sesamum* cp Genome

The slow rate of nucleotide substitution in protein-coding genes is a primary reason for the use of chloroplast genes in plant phylogenetic research at higher taxonomic levels [39], [40]. The nucleotide substitution rates in the intergenic spacer and intron regions are higher than the coding sequence (CDS) regions [41], [42]. Such differences in evolution rates are dependent on the sequence and the gene functions. In addition, several previous studies have reported evolutionary differences in cp DNA sequences related to the structural constraints imposed on the plant cp genomes [43]. Most land plant cp genomes include two identical copies of inverted repeat regions. The frequent intra-chromosomal recombination events between these two IR regions of the cp genome provide selective constraints, both on sequence homogeneity and on structural stability [27], [44]. Therefore, the IR region exhibits slower nucleotide substitution rates in comparison to the SSC and LSC regions.

To address the evolutionary constraint issue in the *Sesamum* cp genome, a series of comparative sequence analyses were conducted using *Sesamum* cp DNA sequences along with the published cp genome sequences of *Olea, Nicotiana* and *Panax* (Table S1, S2, and S3). These three sequences were selected because they belong to the same or closely related taxonomic orders, Lamiales and Solanales. The gene order of the cp genomes was co-linear among these four genera. An alignment of the protein-coding genes, introns, and intergenic spacer regions, along with positional information of the cp genomes for *Sesamum* and three other genera was performed. Of 114 genes, 84 CDS were analyzed. The 30 tRNA genes were excluded in this comparative analysis due to their short length. A total of 110 IGS and 19 intron sequences were also analyzed. First, the sequence comparison data was partitioned into CDS, intron, and IGS regions. The sequence divergence ratios among the three regions (CDS:intron:IGS) were 1:1.3:2.2 between *Sesamum* and *Olea*, 1:1.3:2.3 between *Sesamum* and*Nicotiana*, and 1:1.3:2.1 between *Sesamum* and *Panax* (Table 5). The ratios in these three comparisons are similar. This clearly suggests that the intron sequences have evolved more rapidly than the CDS but slower than the IGS sequences. Second, the sequence comparison data was partitioned into IR, LSC and SSC regions. The sequence divergence ratios among the three regions (IR:LSC:SSC) were 1:4.7:6.9 between *Sesamum* and *Olea*, 1:5.3:6.8 between*Sesamum* and *Nicotiana*, and 1:5.2:6.2 between *Sesamum* and *Panax*. That the ratios are relatively consistent between three different comparisons clearly suggests that the IR regions have evolved much slower than the LSC and SSC regions (Figure 6). The same tendencies are prominent even when comparing the CDS or noncoding sequences for each of the three regions separately. As an example, 19 intron sequences show markedly slow down patterns of base substitutions in IR regions (Figure 3). Furthermore, the Ka/Ks ratio data for the CDS also indicate that the IR region has stronger selection pressures than either the LSC or SSC regions; therefore, these data confirm that positional effects are stronger constraints for sequence evolution than the functional groups of chloroplast genes.

Table 5. Comparisons of protein coding genes (CDS), introns, and intergenic spacers (IGS) at the IR, LSC, and SSC regions of the chloroplast genomes.

[illegible]		[illegible]							[illegible]							[illegible]						
[illegible]		[illegible]	[illegible]	[illegible]	[illegible]	[illegible]	[illegible]	[illegible]	[illegible]	[illegible]	[illegible]	[illegible]	[illegible]	[illegible]	[illegible]	[illegible]	[illegible]	[illegible]	[illegible]	[illegible]	[illegible]	[illegible]
[illegible]	[illegible]	[illegible]	[illegible]	[illegible]	[illegible]	[illegible]	[illegible]	[illegible]	[illegible]	[illegible]	[illegible]	[illegible]	[illegible]	[illegible]	[illegible]	[illegible]	[illegible]	[illegible]	[illegible]	[illegible]	[illegible]	[illegible]
	[illegible]	[illegible]	[illegible]	[illegible]	[illegible]	[illegible]	[illegible]	[illegible]	[illegible]	[illegible]	[illegible]	[illegible]	[illegible]	[illegible]	[illegible]	[illegible]	[illegible]	[illegible]	[illegible]	[illegible]	[illegible]	[illegible]
	[illegible]	[illegible]	[illegible]	[illegible]	[illegible]	[illegible]	[illegible]	[illegible]	[illegible]	[illegible]	[illegible]	[illegible]	[illegible]	[illegible]	[illegible]	[illegible]	[illegible]	[illegible]	[illegible]	[illegible]	[illegible]	[illegible]
	[illegible]	[illegible]	[illegible]	[illegible]	[illegible]	[illegible]	[illegible]	[illegible]	[illegible]	[illegible]	[illegible]	[illegible]	[illegible]	[illegible]	[illegible]	[illegible]	[illegible]	[illegible]	[illegible]	[illegible]	[illegible]	[illegible]
[illegible]	[illegible]	[illegible]	[illegible]	[illegible]	[illegible]	[illegible]	[illegible]	[illegible]	[illegible]	[illegible]	[illegible]	[illegible]	[illegible]	[illegible]	[illegible]	[illegible]	[illegible]	[illegible]	[illegible]	[illegible]	[illegible]	[illegible]
	[illegible]	[illegible]	[illegible]	[illegible]	[illegible]	[illegible]	[illegible]	[illegible]	[illegible]	[illegible]	[illegible]	[illegible]	[illegible]	[illegible]	[illegible]	[illegible]	[illegible]	[illegible]	[illegible]	[illegible]	[illegible]	[illegible]
	[illegible]	[illegible]	[illegible]	[illegible]	[illegible]	[illegible]	[illegible]	[illegible]	[illegible]	[illegible]	[illegible]	[illegible]	[illegible]	[illegible]	[illegible]	[illegible]	[illegible]	[illegible]	[illegible]	[illegible]	[illegible]	[illegible]
	[illegible]	[illegible]	[illegible]	[illegible]	[illegible]	[illegible]	[illegible]	[illegible]	[illegible]	[illegible]	[illegible]	[illegible]	[illegible]	[illegible]	[illegible]	[illegible]	[illegible]	[illegible]	[illegible]	[illegible]	[illegible]	[illegible]
[illegible]	[illegible]	[illegible]	[illegible]	[illegible]	[illegible]	[illegible]	[illegible]	[illegible]	[illegible]	[illegible]	[illegible]	[illegible]	[illegible]	[illegible]	[illegible]	[illegible]	[illegible]	[illegible]	[illegible]	[illegible]	[illegible]	[illegible]
	[illegible]	[illegible]	[illegible]	[illegible]	[illegible]	[illegible]	[illegible]	[illegible]	[illegible]	[illegible]	[illegible]	[illegible]	[illegible]	[illegible]	[illegible]	[illegible]	[illegible]	[illegible]	[illegible]	[illegible]	[illegible]	[illegible]
	[illegible]	[illegible]	[illegible]	[illegible]	[illegible]	[illegible]	[illegible]	[illegible]	[illegible]	[illegible]	[illegible]	[illegible]	[illegible]	[illegible]	[illegible]	[illegible]	[illegible]	[illegible]	[illegible]	[illegible]	[illegible]	[illegible]
	[illegible]	[illegible]	[illegible]	[illegible]	[illegible]	[illegible]	[illegible]	[illegible]	[illegible]	[illegible]	[illegible]	[illegible]	[illegible]	[illegible]	[illegible]	[illegible]	[illegible]	[illegible]	[illegible]	[illegible]	[illegible]	[illegible]
[illegible]		[illegible]	[illegible]	[illegible]	[illegible]	[illegible]	[illegible]	[illegible]	[illegible]	[illegible]	[illegible]	[illegible]	[illegible]	[illegible]	[illegible]	[illegible]	[illegible]	[illegible]	[illegible]	[illegible]	[illegible]	[illegible]

[illegible]

doi:10.1371/journal.pone.0035872.t005

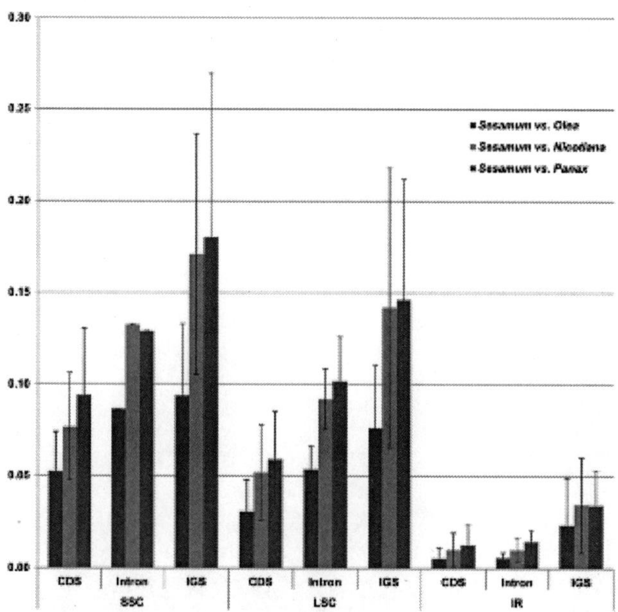

Figure 6. The levels of evolutionary divergences among the SSC, LSC, and IR regions of cp genomes. Y-axis represents the sequence divergences. The IR region evolves slower than the SSC or the LSC regions regardless the CDS, intron and IGS.

doi:10.1371/journal.pone.0035872.g006

Previous research has indicated that in cp genomes, the IR regions are more conserved than the single copy regions [27], [44], [45]. Between two strands of homologous IR sequences, recombination events occur frequently and successive base collection mechanisms break out; therefore, the base substitution rate in the IR region is slower than that of the LSC and SSC regions [27], [44]. In this report, the cp DNA data was partitioned into two different categories: functional constraints and regional constraints (or positional effects). Data indicate that the regional constraints strongly affect the sequence evolution of cp genomes, while the functional constraints weakly affect the sequence evolution (Figures 3 and 5). Fewer indel events also occur in the IR regions than in the LSC or SSC regions [46].

The indel patterns of chloroplast genomes from the three different hierarchical comparisons are summarized in Figure 7. The data suggest that similar indel patterns are observed, regardless of the taxonomic hierarchies. The data also indicate that large indels are relatively rare and that the majority of indels are less than 10 bp in length (Figure 7, Table S4).

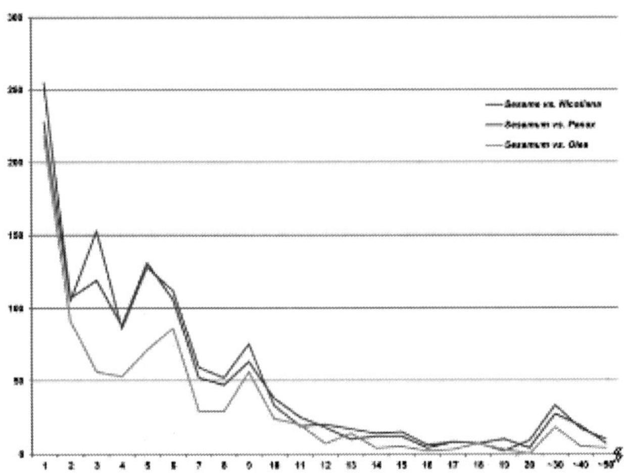

Figure 7. The correlation pattern of indel numbers and indel sizes among three cp genomes. The X-axis and Y-axis represent the indel sizes in base pair and indel numbers, respectively.

doi:10.1371/journal.pone.0035872.g007

Possible Implications for Chloroplast Engineering

The large repeats (26 bp or longer) that exhibited a sequence identity greater than 90% were examined. Many of these repeats contain overlapping components at the same location within the cp genome; therefore, the repeating unit, repeating time and repeating location were determined using REPuter program [47]. Ultimately, seven total repeats were identified and localized (Figure 1). These included two direct repeats, two direct inverted repeats, and three palindromic dispersed repeats as possible gene introduction sites (Table 4). These genes are repeated two to four times. Repeat no. 3, which occurs in three different regions of the cp genome including the two IR and the SSC regions, may have limited utility as a site-specific recombination site. Repeats No. 4 and 5 also show similar challenges for use as vector sites. Two different palindromic repeats are located in the intergenic spacers of the LSC region between *psbT* and *psbN* and between *petN* and *psbM*. These repeats may be useful for the development of site-specific recombination sites for foreign gene cassettes. Two additional useful direct repeats are located on the CDS of the *ycf2* gene in the IR regions. In *Sesamum*, this *ycf2* gene is 6,294 bp long, has an unknown function, and is tolerant of large indel mutations. One direct repeat is especially important because it is 81 bp long and will easily accommodate site-specific recombination. As a result, the two direct and two palindromic repeats present possible foreign gene introduction sites. Three of the seven large repeat loci in the *Sesamum* cp genome are also conserved in the *Olea* cp genome.

In recent years, the universal vector located in the *trnA/trnI* IGS region has been used as a gene introduction site for cp genome engineering [48]. However, it can only be used for plants that are closely related and show high levels of genome sequence homology. This vector has limited utility if the sequences are substantially different; therefore, a species-specific cp vector is expected to be more reliable for plant gene transformation [20]. The complete cp genome sequences are required for the development of site-specific chloroplast vector sites. The genes related to lipid biosynthesis will be primary target genes for alteration in *Sesamum*. The *ACP desaturase* (*SAD*) and *FAD2* genes have been used to produce sunflower [49], [50], [51] and soybean [52], [53] transgenic plants that exhibit oil modification. These genes could be similarly modified in *Sesamum*. The two genes, along with other genes involved in lipid biosynthesis, can be engineered in a single cassette for introduction into the *Sesamum* cp genome. The direct or palindromic repeat sites of the *Sesamum* cp genome represent potential cassette introduction sites that could be used in the development of a sesame-specific chloroplast vector, similar to the *trnA/trnI* flanking sequences used in the universal cp vectors for Solanaceous plants.

Utility of Repeat Units and cp SSRs

The function and origin of SSRs within the chloroplast genome are not yet fully understood; however, SSR loci are typically present in plant cp genomes and can provide useful information concerning plant population genetics [54], [55]. The presence of SSRs in cp genomes was initially reported in *Pinus radiata* and *Oryza sativa* [34], [56], [57]. Later, Kim and Lee also reported 18 SSR loci and 29 SSR loci in the cp genomes of *Panax* and *Nicotiana*, respectively[58].

Twenty-eight SSR loci were identified in the *Sesamum* cp genome (Table 3). Of these, 21 are homopolymers, four are di-polymers, and three are tri-polymers. Of the homopolymer loci, 18 are composed of A or T multiples, while only three are composed of C or G multiples. All of the di-polymer loci are composed of multiples of AT or TA. Three SSR loci were identical to loci in the *Olea* cp genome. Length variations in SSR loci serve as useful markers for identifying varieties of crops and population genetics [34], [59], [60], [61]. *Sesamum indicum*, which is widely cultivated, has nearly 3,000 cultivars. Using breeding and selection approaches, over 38,000 genetic lines have been developed (United States Department of Agriculture, 2010). Cultivars are distinguished by capsule numbers per node; locule numbers within a capsule; stem branching patterns; seed shapes and colors; flower colors; leaf margin shapes; plant height; trichomes on the fruit, stem, and leaf; fruit maturation; and more [62], [63], [64]; however, many cultivars are difficult to distinguish using these morphological characters. If the cp SSR information is compiled, these SSR loci can provide useful identification tools for some of these cultivars. The complete cp DNA sequences of *Sesamum indicum*, as well as the SSR loci information, provide invaluable sources for developing primers to study specific SSR loci.

Phylogenetic Position and Origin of *Sesamum* (Pedaliaceae)

Complete cp genome sequences provide rich sources of phylogenetic information. Therefore, several recent phylogenetic studies based on the complete cp genome sequences are addressed successfully for the phylogenetic issues of angiosperm [12], [13]. These genome based analyses across whole angiosperm lineages usually used 61–81 protein coding sequences to assembling the data matrix because of the missing genes in some lineages. Previous genome scale analysis included the 18 complete cp genomes from asterid lineages[12], [65]. In this study, however, we aligned 83 genes from 32 complete cp genome sequences which representing 10 families and 5 orders of asterids. Therefore, our analysis represents the most comprehensive data from asterids. Our phylogenetic tree almost identical to the Angiosperm Phylogeny Group (APG: http://

www.mobot.org/mobot/research/apweb) tree and represent the subset of the APG tree. *Sesamum* (Pedaliaceae) form a sister group to the *Olea*and *Jasminum* (Oleaceae) clade (Figure 5). Oleaceae usually positioned as a basal sister family to other Lamiales families [38]. Therefore, our complete cp genome sequences of *Sesamum*represent the core lineage of Lamiales families. The data will be served as a reference sequence for the future genome scale phylogenetic study of this problematic group.

Two major lineages of asterids, asterid I and II, diverged between 114.3±6.7 million years ago (mya) in our tree (Figure 5). This time estimation is highly comparable to the 117–107 mya of the previous reports [35], [66]. Three major orders (Lamiales, Solanales and Gentianales) of asterid I lineages were diversified between 104.2–98.8 mya and it also comparable to the previous estimations of 95±12 mya [67], [68]. Finally, our tree also suggests that the splitting time of Oleaceae (represented by *Jasminum* and *Olea*) and the core Lamiales (represented by*Sesamum*) were approximately 70.1±5.5 mya (Figure 5). It corresponds to the K-T boundary of geological time scale.

Materials and Methods

Plants Materials and cpDNA Isolation

Thirty *Sesamum indicum* L. cv. Ansanggae (a black-seeded cultivar) plants were cultivated from seeds originating from a single seed pod of a mother plant. Approximately 100 grams of fresh leaves were harvested from the 30 mature individuals, and two voucher specimens were deposited in the Korea University Herbarium (KUS). To remove starch and sugar from the cells, the fresh leaves were kept in the dark for 48 hrs at 0°C prior to organelle isolation. The leaf tissues were ground using a conventional blender and Sorbitol/TE isolation buffer (0.35 M sorbitol, 50 mM Tris-HCl, 5 mM EDTA, pH 8.0, 0.1% BSA, 0.1% 2-mercaptoethanol). The homogenate is filtered through two layers of miracloth (Calbiochem) and centrifuged at 1,000 g for 15 min at 4°C. The intact cp organelles were purified using sucrose step gradient centrifugation [69]. High purity cp organelles were obtained from the 52–30% sucrose interface. Cp organelles were collected from a total of 12 sucrose gradient tubes in 50 ml volumes. After the careful washing the cp organelles in wash buffer (0.35 M sorbitol, 50 mM Tris-HCl, 5 mM EDTA, pH 8.0, 0.1% BSA), cpDNA was isolated from lysed chloroplasts using ultracentrifugation in a cesium chloride/ethidium bromide gradient. Impurities were removed by dialysis. CpDNAs (Plant DNA Bank of Korea accession number 1996-0001) were quantified using NanoDrop spectrophotometers (Thermo Scientific, Nanodrop 2000), and the cpDNA quality was analyzed on a 1% agarose gel following *BamH*I and *Sac*I restriction enzyme digestion.

PCR Amplification and Sequencing

Chloroplast DNA sequences were analyzed using the GS-FLX pyrosequencing method [70] and the Genome Sequencer FLX system (Roche, Basal, Switzerland). A total of 133,533 reads, with an average read length of 236 bp, were analyzed to generate 31,540,819 bp of sequence. Because of the contamination of nuclear and mitochondrial DNA in cpDNA, we filtered all reads by extensive BLAST searches using the reference cpDNA sequences from *Panax* [58]. The filtered sequences were assembled using the Newbler program (Roche Diagnostics Company). The combination of the high purity cp organelle isolation procedure and extremely high sequence coverage enabled the assembly of contigs that nearly spanned the entire cp genome. Using BLAST comparisons (BLASTN, PHI-BLAST and BLASTX), we identified 155 contigs that is cp DNA sequences. Of these, three large contigs (85,165 bp, 25,137 bp, 17,877 bp) corresponded to the LSC, IR and SSC regions of the *Sesamum* cp genome, respectively. An additional 152 short contigs were included in the three large contigs. The total length of the contigs was 257,427 bp with an average contig size of 1,660 bp. Gaps between the three large contigs were filled via direct sequencing of PCR products amplified using primers that were complementary to the end sequences of each contig. The amplified regions corresponded to the IR/SSC and IR/LSC boundaries. The sequenced fragments were assembled using Sequencher 4.8 (Gene Code Corporation, Ann Arbor, MI, USA).

Chloroplast Gene Annotation and Sequence Analyses

Gene annotations and comparative analysis were performed using the BLAST (BLASTN, PHI-BLAST, BLASTX) ORF finder program from the National Center for Biotechnology Information (NCBI) and DOGMA [71]. The nomenclature of cp gene is follows the Chloroplast Genome Database (http://chloroplast.cbio.psu.edu). Codon usage and A-T contents were analyzed using MEGA4 (version 4.1) [72]. Repeating sequences were analyzed using REPuter [47] and further analyzed with the Tandem Repeats Finder, ver. 4.0 [73]. Twenty-eight SSR loci were identified in the *Sesamum* cp genome (Table 3). All SSR regions are PCR amplified and re-sequenced manually in order to prevent the error in pyrosequencing procedure. For sequence comparisons, the gene, intron, and gene spacer regions from the cp genomes of different species were aligned using Clustal X 2.0 [74] and adjusted by hand. Several spacer regions were aligned using MUSCLE [75]. mVISTA were used to compare similarities among different chloroplast genomes [76]. Nucleotide diversity and Ka/Ks value were analyzed using DnaSP (version 4.50) [77]. Secondary structure predicted by mFOLD [78] and TRNAscan-SE [79].

Phylogenetic Analysis

Thirty-two complete cp DNA sequences representing the asteroid lineage of angiosperm were obtained from NCBI databases (Table S5). For the phylogenetic analysis, 83 gene sequences were initially aligned using the Clustal algorithm [74] and then realigned by the MUSCLE program [75]. Maximum likelihood (ML) analysis was performed using PAUP version 4.0b10 [80]with Modeltest [81]. The GTR+G+I base substitution model was adopted. The Bayesian supporting values of all internal nodes were also calculated under the options of rep = 250,000, lset nst = 6, rates = gamma, basefreq = estimate and burnin = 5000. Molecular time estimation was done using the r8s program [82] implementing semiparametric rate smoothing by penalized likelihood.

AUTHOR CONTRIBUTIONS

Conceived and designed the experiments: KJK. Performed the experiments: DKY. Analyzed the data: DKY. Contributed reagents/materials/analysis tools: KJK. Wrote the paper: DKY KJK.

REFERENCES

1. Bedigian D (2003) Evolution of sesame revisited: domestication, diversity and prospects. Genet Resour Crop Evol 50: 773–778.
2. Bedigian D, Seigler DS, Harlan JR (1985) Sesamin, sesamolin and the origin of sesame. Biochem System Ecol 13: 133–139.
3. Chen ECF, Tai SSK, Peng CC, Tzen JTC (1998) Identification of three novel unique proteins in seed oil bodies of sesame. Plant Cell Physiol 39: 935–941.
4. Pham TD, Bui TM, Werlemark G, Bui TC, Merker A, et al. (2009) A study of genetic diversity of sesame (*Sesamum indicum* L.) in Vietnam and Cambodia estimated by RAPD markers. Genet Resour Crop Evol 56: 679–690.
5. Parsaeian M, Mirlohi A, Saeidi G (2011) Study of genetic variation in sesame (*Sesamum indicum* L.) using agro-morphological traits and ISSR markers. Genetika 47: 359–367.
6. Wei WL, Qi XQ, Wang LH, Zhang YX, Hua W, et al. (2011) Characterization of the sesame (*Sesamum indicum* L.) global transcriptome using Illumina paired-end sequencing and development of EST-SSR markers. BMC Genomics 12: 451.
7. Kim DH, Zur G, Danin-Poleg Y, Lee SW, Shim KB, et al. (2002) Genetic relationships of sesame germplasm collection as revealed by inter-simple sequence repeats. Plant Breeding 121: 259–262.
8. Shinozaki K, Ohme M, Tanaka M, Wakasugi T, Hayashida N, et al. (1986) The complete nucleotide sequence of tobacco chloroplast genome: its gene organization and expression. EMBO J 5: 2043–2049.
9. Shinozaki K, Hayashida N, Sugiura M (1988) Nicotiana chloroplast genes for components of the photosynthetic apparatus. Photosynth Res 18: 7–31.
10. Raubeson LA, Jansen RK (2005) Chloroplast genomes of plants. In: Henry R, editor. Diversity and evolution of plants-genotypic variation in higher plants. Oxfordshire: CABI Publishing. pp. 45–68.
11. Downie SR, Palmer JD (1992) Use of chloroplast DNA rearrangements

in reconstructing plant phylogeny. In: Soltis PS, Soltis DE, Doyle JJ, editors. Molecular Systematics of Plants New York: Chapman and Hall. pp. 14–35.

12. Jansen RK, Cai Z, Raubeson LA, Daniell H, Depamphilis CW, et al. (2007) Analysis of 81 genes from 64 plastid genomes resolves relationships in angiosperms and identifies genome-scale evolutionary patterns. Proc Natl Acad Sci USA 104: 19369–19374.
13. Moore MJ, Bell CD, Soltis PS, Soltis DE (2007) Using plastid genome-scale data to resolve enigmatic relationships among basal angiosperms. Proc Natl Acad Sci USA 104: 19363–19368.
14. Sugiura M (1992) The chloroplast genome. Plant Mol Biol 19: 149–168.
15. Ogihara Y, Terachi T, Sasakuma T (1988) Intramolecular recombination of chloroplast genome mediated by short direct-repeat sequences in wheat species. Proc Natl Acad Sci USA 85: 8573–8577.
16. Jansen RK, Palmer JD (1987) A chloroplast DNA inversion marks an ancient evolutionary split in the sunflower family (Asteraceae). Proc Natl Acad Sci USA 84: 5818–5822.
17. Graham SW, Reeves PA, Burns ACE, Olmstead RG (2000) Microstructural changes in noncoding chloroplast DNA: Interpretation, evolution, and utility of indels and inversions in basal angiosperm phylogenetic inference. Int J Plant Sci 161: S83–S96.
18. Matsuoka Y, Yamazaki Y, Ogihara Y, Tsunewaki K (2002) Whole chloroplast genome comparison of rice, maize, and wheat: Implications for chloroplast gene diversification and phylogeny of cereals. Mol Biol Evol 19: 2084–2091.
19. De las Rivas J, Lozano JJ, Ortiz AR (2002) Comparative analysis of chloroplast genomes: functional annotation, genome-based phylogeny, and deduced evolutionary patterns. Genome Res 12: 567–583.
20. Verma D, Daniell H (2007) Chloroplast vector systems for biotechnology applications. Plant Physiol 145: 1129–1143.
21. Kumar S, Daniell H (2004) Engineering the chloroplast genome for hyperexpression of human therapeutic proteins and vaccine antigens. Methods Mol Biol 267: 365–383.
22. Kumar S, Dhingra A, Daniell H (2004) Plastid-expressed betaine aldehyde dehydrogenase gene in carrot cultured cells, roots, and leaves confer enhanced salt tolerance. Plant Physiol 136: 2843–2854.
23. Daniell H, Lee SB, Panchal T, Wiebe PO (2001) Expression of the native cholera toxin B subunit gene and assembly as functional oligomers in transgenic tobacco chloroplasts. J Mol Biol 311: 1001–1009.

24. Ogihara Y, Isono K, Kojima T, Endo A, Hanaoka M, et al. (2002) Structural features of a wheat plastome as revealed by complete sequencing of chloroplast DNA. Mol Genet Genomics 266: 740–746.

25. Kato T, Kaneko T, Sato S, Nakamura Y, Tabata S (2000) Complete structure of the chloroplast genome of a legume, *Lotus japonica*. DNA Res 7: 323–330.

26. Maier RM, Neckermann K, Igloi GL, Kossel H (1995) Complete sequence of the maize chloroplast genome - gene content, hotspots of divergence and fine-tuning of genetic information by transcript editing. J Mol Biol 251: 614–628.

27. Palmer JD (1991) Plastid chromosomes: structure and evolution. In: Vasil IK, Bogorad L, editors. Cell Culture and Somatic Cell Genetics in Plants, Vol 7A, The Molecular Biology of Plastids. San Diego: Academic Press. pp. 5–53.

28. Hiratsuka J, Shimada H, Whittier R, Ishibashi T, Sakamoto M, et al. (1989) The complete sequence of the rice (*Oryza sativa*) chloroplast genome - intermolecular recombination between distinct tRNA genes accounts for a major plastid DNA inversion during the evolution of the cereals. Mol Gen Genet 217: 185–194.

29. Catalano SA, Saidman BO, Vilardi JC (2009) Evolution of small inversions in chloroplast genome: a case study from a recurrent inversion in angiosperms. Cladistics 25: 93–104.

30. Bain JF, Jansen RK (2006) A chloroplast DNA hairpin structure provides useful phylogenetic data within tribe Senecioneae (Asteraceae). Can J Bot 84: 862–868.

31. Kim KJ, Lee HL (2005) Widespread occurrence of small inversions in the chloroplast genomes of land plants. Mol Cells 19: 104–113.

32. Mes THM, Kuperus P, Kirschner J, Stepanek J, Oosterveld P, et al. (2000) Hairpins involving both inverted and direct repeats are associated with homoplasious indels in non-coding chloroplast DNA of *Taraxacum* (Lactuceae : Asteraceae). Genome 43: 634–641.

33. Grassi F, Labra M, Scienza A, Imazio S (2002) Chloroplast SSR markers to assess DNA diversity in wild and cultivated grapevines. Vitis 41: 157–158.

34. Powell W, Morgante M, Mcdevitt R, Vendramin GG, Rafaslki JA (1995) Polymorphic simple sequence repeat regions in chloroplast genomes: applications to the population genetics of pines. Proc Natl Acad Sci USA 92: 7759–7763.

35. Bremer K, Friis EM, Bremer B (2004) Molecular phylogenetic dating of asterid flowering plants shows early Cretaceous diversification. Syst

Biol 53: 496–505.

36. Wikstrom N, Savolainen V, Chase MW (2001) Evolution of the angiosperms: calibrating the family tree. Proc Roy Soc B Biol Sci 268: 2211–2220.
37. Yue F, Zhang M, Tang JJ (2007) Phylogenetic reconstruction from transpositions. BMC Genomics 9: S15.
38. Lee HL, Jansen RK, Chumley TW, Kim KJ (2007) Gene relocations within chloroplast genomes of *Jasminum* and *Menodora* (Oleaceae) are due to multiple overlapping inversions. Mol Biol Evol 24: 1161–1180.
39. Curtis SE, Clegg MT (1984) Molecular Evolution of Chloroplast DNA Sequences. Mol Bio Evol 1: 291–301.
40. Wolfe KH, Li WH, Sharp PM (1987) Rates of nucleotide substitution vary greatly among plant mitochondrial, chloroplast and nuclear DNAs. Proc Natl Acad Sci USA 84: 9054–905.
41. Morton BR, Clegg MT (1995) Neighboring base composition in strongly correlated with base substitution in a region of the chloroplast genome. J Mol Evol 41: 597–603.
42. Clegg MT, Gaut BS, Learn GH, Morton BR (1994) Rates and patterns of chloroplast DNA evolution. Proc Natl Acad Sci USA 91: 6795–6801.
43. Perry AS, Wolfe KH (2002) Nucleotide substitution rates in legume chloroplast DNA depend on the presence of the inverted repeat. J Mol Evol 55: 501–508.
44. Wolfe KH, Gouy ML, Yang YW, Sharp PM, Li WH (1989) Date of the monocot-dicot divergence estimated from chloroplast DNA sequence data. Proc Natl Acad Sci USA 86: 6201–6205.
45. Kim YK, Park CW, Kim KJ (2009) Complete chloroplast DNA sequence from a Korean endemic genus, *Megaleranthis saniculifolia,* and its evolutionary implications. Mol Cells 27: 365–381.
46. Yamane K, Yano K, Kawahara T (2006) Pattern and rate of indel evolution inferred from whole chloroplast intergenic regions in sugarcane, maize and rice. DNA Res 13: 197–204.
47. Kurtz S, Choudhuri JV, Ohlebusch E, Schleiermacher C, Stoye J, et al. (2001) REPuter: the manifold applications of repeat analysis on a genomic scale. Nucl Acids Res 29: 4633–4642.
48. Daniell H, Datta R, Varma S, Gray S, Lee SB (1998) Containment of herbicide resistance through genetic engineering of the chloroplast genome. Nat Biotechnol 16: 345–348.
49. Weber S, Friedt W, Landes N, Molinier J, Himber C, et al. (2003) Improved Agrobacterium-mediated transformation of sunflower (*Helianthus annuus* L.): assessment of macerating enzymes and sonication.

Plant Cell Rep 21: 475–482.

50. Rousselin P, Molinier J, Himber C, Schontz D, Prieto-Dapena P, et al. (2002) Modification of sunflower oil quality by seed-specific expression of a heterologous Δ9-stearoyl-(acyl carrier protein) desaturase gene. Plant Breeding 121: 108–116.

51. Martinez-Rivas JM, Garcia-Diaz MT, Mancha M (2000) Temperature and oxygen regulation of microsomal oleate desaturase (FAD2) from sunflower. Biochem Soc Trans 28: 890–892.

52. Clemente TE, Cahoon EB (2009) Soybean oil: genetic approaches for modification of functionality and total content. Plant Physiol 151: 1030–1040.

53. Heppard EP, Kinney AJ, Stecca KL, Miao GH (1996) Developmental and growth temperature regulation of two different microsomal ω-6 desaturase genes in soybeans. Plant Physiol 110: 311–319.

54. Echt CS, DeVerno LL, Anzidei M, Vendramin GG (1998) Chloroplast microsatellites reveal population genetic diversity in red pine, *Pinus resinosa* Ait. Mol Ecol 7: 307–316.

55. Powell W, Morgante M, Andre C, Mcnicol JW, Machray GC, et al. (1995) Hypervariable microsatellites provide a general source of polymorphic DNA markers for the chloroplast genome. Curr Biol 5: 1023–1029.

56. Cato SA, Richardson TE (1996) Inter- and intraspecific polymorphism at chloroplast SSR loci and the inheritance of plastids in *Pinus radiata* (D. Don.). Theor Appl Genet 93: 587–592.

57. Provan J, Corbett G, Waugh R, McNicol JW, Morgante M, et al. (1996) DNA fingerprints of rice (*Oryza sativa*) obtained from hypervariable chloroplast simple sequence repeats. Proc Roy Soc Lond B Bio 263: 1275–1281.

58. Kim K-J, Lee H-L (2004) Complete chloroplast genome sequences from Korean ginseng (*Panax schinseng* Nees) and comparative analysis of sequence evolution among 17 vascular plants. DNA Res 11: 247–261.

59. Garris AJ, Tai TH, Coburn J, Kresovich S, McCouch S (2005) Genetic structure and diversity in *Oryza sativa* L. Genetics 169: 1631–1638.

60. Xu DH, Abe J, Gai JY, Shimamoto Y (2002) Diversity of chloroplast DNA SSRs in wild and cultivated soybeans: evidence for multiple origins of cultivated soybean. Theor Appl Genet 105: 645–653.

61. Bryan GJ, McNicoll J, Ramsay G, Meyer RC, De Jong WS (1999) Polymorphic simple sequence repeat markers in chloroplast genomes of Solanaceous plants. Theor Appl Genet 99: 859–867.

62. Adebowale AA, Sanni SA, Falore OA (2011) Varietal difference in the

physical properties and proximate composition of elite sesame seeds. World J Agric Sci 7: 42–46.

63. Pham TD, Nguyen TDT, Carlsson AS, Bui TM (2010) Morphological evaluation of sesame (*Sesamum indicum* L.) varieties from different origins. Australian J Crop Sci 4: 498–504.
64. Ong'injo EO, Ayiecho PO (2009) Genotypic variability in sesame mutant lines in Kenya. African Crop Sci J 17: 101–107.
65. Leebens-Mack J, Raubeson LA, Cui LY, Kuehl JV, Fourcade MH, et al. (2005) Identifying the basal angiosperm node in chloroplast genome phylogenies: sampling one's way out of the felsenstein zone. Mol Biol Evol 22: 1948–1963.
66. Soltis DE, Bell CD, Kim S, Soltis PS (2008) The origin and early evolution of angiosperms. Ann New York Acad Sci 1133: 3–25.
67. Janssens SB, Knox EB, Huysmans S, Smets EF, Merckx VSFT (2009) Rapid radiation of*Impatiens* (Balsaminaceae) during Pliocene and Pleistocene: result of a global climate change. Mol Phylogenet Evol 52: 806–824.
68. Magallón S, Castillo A (2009) Angiosperm diversification through time. American J Bot 96: 349–365.
69. Palmer JD (1986) Isolation and structural analysis of chloroplast DNA. In: Weissbach A, Weissbach H, editors. Methods in Enzymology, Vol 118. New York: Academic Press. pp. 167–186.
70. Margulies M, Egholm M, Altman WE, Attiya S, Bader JS, et al. (2005) Genome sequencing in microfabricated high-density picolitre reactors. Nature 437: 376–380.
71. Wyman SK, Jansen RK, Boore JL (2004) Automatic annotation of organelle genomes with DOGMA. Bioinformatics 20: 3252–3255.
72. Tamura K, Dudley J, Nei M, Kumar S (2007) MEGA4: molecular evolutionary genetics analysis (MEGA) software version 4.0. Mol Bio Evol 24: 1596–1599.
73. Benson G (1999) Tandem repeats finder: a program to analyze DNA sequences. Nucleic Acids Res 27: 573–580.
74. Thompson JD, Higgins DG, Gibson TJ (1994) CLUSTAL W: improving the sensitivity of progressive multiple sequence alignment through sequence weighting position-specific gap penalties and weight matrix choice. Nucleic Acids Res 22: 4673–4680.
75. Edgar RC (2004) MUSCLE: a multiple sequence alignment method with reduced time and space complexity. BMC Bioinformatics 5: 1–19.
76. Mayor C, Brudno M, Schwartz JR, Poliakov A, Rubin EM, et al. (2000) VISTA: visualizing global DNA sequence alignments of arbitrary

length. Bioinformatics 16: 1046–1047.

77. Librado P, Rozas J (2009) DnaSP v5: a software for comprehensive analysis of DNA polymorphism data. Bioinformatics 25: 1451–1452.
78. Zuker M (2003) Mfold web server for nucleic acid folding and hybridization prediction. Nucleic Acids Res 31: 3406–3415.
79. Lowe TM, Eddy SR (1997) tRNAscan-SE: A program for improved detection of transfer RNA genes in genomic sequence. Nucl Acids Res 25: 955–964.
80. Swofford DL (2002) PAUP*: phylogenetic analysis using parsimony (*and other methods). Version 4. Sunderland, Massachusetts: Sinauer Associates.
81. Posada D, Crandall KA (1998) MODELTEST: testing the model of DNA substitution. Bioinformatics 14: 817–818.
82. Sanderson MJ (2003) r8s: inferring absolute rates of molecular evolution and divergence times in the absence of a molecular clock. Bioinformatics 19: 301–302.

Chapter 6

GENETICALLY MODIFIED CROPS AND FOOD SECURITY

[1]Matin Qaim, [2]Shahzad Kouser

[1]Department of Agricultural Economics and Rural Development, Georg-August-University of Goettingen, Goettingen, Germany

[2]Institute of Agricultural and Resource Economics, University of Agriculture, Faisalabad, Pakistan

ABSTRACT

The role of genetically modified (GM) crops for food security is the subject of public controversy. GM crops could contribute to food production increases and higher food availability. There may also be impacts on food quality and nutrient composition. Finally, growing GM crops may influence farmers' income and thus their economic access to food. Smallholder farmers make up a large proportion of the undernourished people worldwide. Our study focuses on this latter aspect and provides the first *ex post* analysis of food security impacts of GM crops at the micro level. We use comprehensive panel data collected over several years from farm households in India, where insect-resistant GM cotton has been widely adopted. Controlling for other factors, the adoption of GM cotton has significantly improved calorie consumption and dietary quality, resulting from increased family incomes. This technology has reduced food insecurity by 15–20% among cotton-producing households. GM crops alone will not solve the hunger problem, but they can be an important component in a broader food security strategy.

INTRODUCTION

Food security exists when all people have physical and economic access to sufficient, safe, and nutritious food. Unfortunately, food security does not exist for a significant proportion of the world population. Around 900 million people are undernourished, meaning that they are undersupplied with calories [1]. Many more suffer from specific nutritional deficiencies, often related to insufficient intake of micronutrients. Eradicating hunger is a central part of the United Nations' Millennium Development Goals [2]. But how to achieve this goal is debated controversially. Genetically modified (GM) crops are sometimes mentioned in this connection. Some see the development and use of GM crops as key to reduce hunger [3], [4], while others consider this technology as a further risk to food security [5], [6]. Solid empirical evidence to support either of these views is thin.

There are three possible pathways how GM crops could impact food security. First, GM crops could contribute to food production increases and thus improve the availability of food at global and local levels. Second, GM crops could affect food safety and food quality. Third, GM crops could influence the economic and social situation of farmers, thus improving or worsening their economic access to food. This latter aspect is of particular importance given that an estimated 50% of all undernourished people worldwide are small-scale farmers in developing countries [7].

In regard to the first pathway, GM technologies could make food crops higher yielding and more robust to biotic and abiotic stresses [8], [9]. This could stabilize and increase food supplies, which is important against the background of increasing food demand, climate change, and land and water scarcity. In 2012,

170 million hectares (ha) - around 12% of the global arable land - were planted with GM crops, such as soybean, corn, cotton, and canola [10], but most of these crops were not grown primarily for direct food use. While agricultural commodity prices would be higher without the productivity gains from GM technology [11], impacts on food availability could be bigger if more GM food crops were commercialized. Lack of public acceptance is one of the main reasons why this has not yet happened more widely [12].

Concerning the second pathway, crops with new traits can be associated with food safety risks, which have to be assessed and managed case by case. But such risks are not specific to GM crops. Long-term research confirms that GM technology is not *per se* more risky than conventional plant breeding technologies [13]. On the other hand, GM technology can help to breed food crops with higher contents of micronutrients; a case in point is Golden Rice with provitamin A in the grain [14]. Such GM crops have not yet been commercialized. Projections show that they could reduce nutritional deficiencies among the poor, entailing sizeable positive health effects [15], [16].

The third pathway relates to GM crop use by smallholder farmers in developing countries. Half of the global GM crop area is located in developing countries, but much of this refers to large farms in countries of South America. One notable exception is *Bacillus thuringiensis* (Bt) cotton, which is grown by around 15 million smallholders in India, China, Pakistan, and a few other developing countries [10]. Bt cotton provides resistance to important insect pests, especially cotton bollworms. Several studies have shown that Bt cotton adoption reduces chemical pesticide use and increases yields in farmers' fields [17]–[20]. There are also a few studies that have shown that these benefits are associated with increases in farm household income and living standard [21]–[23]. Higher incomes are generally expected to cause increases in food consumption in poor farm households. On the other hand, cotton is a non-food cash crop, so that the nutrition impact is uncertain.

Here we address this question and analyze the impact of Bt cotton adoption on calorie consumption and dietary quality in India. Bt cotton was first commercialized in India in 2002. In 2012, over 7 million farmers had adopted this technology on 10.8 million ha - equivalent to 93% of the country's total cotton area [10]. For the analysis, we carried out a household survey and collected comprehensive data over a period of several years. This is the first *ex post* study that analyzes food security effects of Bt cotton or any other GM crop with micro level data.

Materials and Methods

Ethics Statement

Our study builds on data from a socioeconomic survey of farm households in India. Details of this survey are explained further below. The institutional review board of the University of Goettingen only reviews clinical research; our study cannot be classified as clinical research. We consulted with the Head of the Research Department of the University of Goettingen, who confirmed that there is no institutional review board at our University that would require a review of such survey-based socioeconomic research.

Farm Household Survey

We carried out a panel survey of Indian cotton farm households in four rounds between 2002 and 2008. We used a multistage sampling procedure. Four states were purposively selected, namely Maharashtra, Karnataka, Andhra Pradesh, and Tamil Nadu. These four states cover a wide variety of different cotton-growing situations, and they produce 60% of all cotton in central and southern India [23]. In these four states, we randomly selected 10 cotton-growing districts and 58 villages, using a combination of census data and agricultural production statistics [18],[19], [23]. Within each village, we randomly selected farm households from complete lists of cotton producers. Sample households were visited individually, and the household head was taken through a face-to-face interview, for which we used a structured questionnaire. The questionnaire covered a wide array of agricultural and socioeconomic information, such as input-output details in cotton production, technology adoption, other income sources, and household living standards. The interviews were carried out in local languages by a small team of enumerators, who were trained and supervised by the researchers.

Prior to starting each interview, the study objective was explained. We also clarified that the data collected would be treated confidentially, analyzed anonymously, and be used for research purposes only. Based on this, the interviewees were asked for their verbal informed consent to participate. We decided not ask for written consent, because the interviews were not associated with any risk for participants. Furthermore, many of the sample farmers had relatively low educational backgrounds and were not used to formal paperwork. Very few households did not agree to participate; they were replaced with other randomly selected households in the same villages.

The first-round survey interviews took place in early 2003, shortly after the cotton harvest for the 2002 season was completed. The same survey was repeated at two-year intervals in early 2005 (referring to the 2004 cotton season), early 2007 (referring to the 2006 season), and early 2009 (referring to the 2008 season). In total, 533 households were interviewed during the 7-year period. Most of these households were visited in several rounds. The total sample consists of 1431 household observations (Table 1). In 2002, the proportion of Bt adopters was still relatively small, but it increased rapidly in the following years. By 2008, 99% of the sample households had adopted this technology. To our knowledge, this is the only longer-term panel survey of Bt cotton farm households in a developing country (the data set with the variables used in this article is available as Data S1).

Table 1. Number of farm households sampled in India in four survey rounds.

Farm households	2002	2004	2006	2008	Total
Adopters of Bt	131	246	333	375	1085
Non-adopters of Bt	210	117	14	5	346
Total	341	363	347	380	1431

doi:10.1371/journal.pone.0064879.t001

doi:10.1371/journal.pone.0064879.t001

Calorie Consumption Data

The survey questionnaire included a detailed food consumption recall, which is a common tool to assess food security at the household level [24]. For a 30-day recall period, households were asked about the quantity consumed of different food items and the corresponding monetary value. The questions covered food consumed from own production, market purchases, gifts, and transfers.

The quantity data for the different food items were converted to calories consumed by using calorie conversion factors for India [25], [26]. The total household calorie consumption from the 30-day recall was then divided by 30 to obtain a calorie value per day. Taking into account the age and gender structure of households, as well as physical activity levels of household members, the number of adult equivalents (AE) was calculated for each household. Male adults involved in farming count as 1.0 AE, female adults involved in farming as 0.8 AE. Male and female adults with lower physical activity levels count as 0.8 and 0.7, respectively. For children and adolescents, appropriate adjustments were made [25]–[27]. The daily household calorie consumption was divided by the number of AE in a household to obtain the calories consumed per AE and day.

Values for minimum dietary energy requirements found in the literature vary, which is due to several reasons [24]. Values stated per capita are lower than those stated per AE, because children have lower calorie requirements than adults. Moreover, not all studies take physical activity levels into account already in the AE calculations, as we do. The average daily calorie requirement for a moderately active AE in India is 2875 kcal/day [25]. According to the World Health Organization, a safe minimum daily intake should not fall below 80% of the calorie requirement, meaning 2300 kcal per AE. Minimum values around 2300 kcal per day for adult men are also found in other studies [28]. Based on this, we take 2300 kcal per AE as the threshold, that is, households with daily calorie consumption below 2300 kcal per AE are considered food insecure.

Most of the calories consumed in rural India are from cereals such as wheat, rice, millet, and sorghum that are rich in carbohydrates but less nutritious in terms of protein and micronutrient contents. Hence, in addition to total calories consumed we calculated the number of calories consumed from more nutritious foods to assess dietary quality. In the category "more nutritious foods", we include pulses, fruits, vegetables, and all animal products (i.e., milk, milk products, meat, fish, and eggs). Recent research suggests that the share of calories consumed from higher value, non-staple foods can also be used as an indicator of nutritional sufficiency [29]. The reason is that poor and undernourished households will largely choose foods that are the cheapest available sources of calories, namely cereals in the context of rural India. Only when they have surpassed subsistence, consumers will begin to substitute towards foods that are more expensive sources of calories [29].

It should be mentioned that food consumption data from household surveys may not provide very accurate data to measure nutritional status [24], [30]. Sometimes, consumption data overestimate calorie intakes, because food losses, waste, and other uses within the household cannot be properly accounted for. However, this limitation applies to both adopters and non-adopters of Bt, so that the comparison between Bt and non-Bt, which is relevant for the impact assessment, is unaffected.

Regression Models

To estimate the impact of Bt cotton adoption on calorie consumption, we regress total daily calorie consumption per AE on Bt adoption, measured as the number of hectares of Bt cotton grown by a household in a particular year. Since Bt adoption increases farm profits and household incomes [23], we expect a positive and significant treatment effect. However, calorie consumption is also influenced by other factors that need to be controlled for. We control for education of the household head (measured in terms of the number of years of schooling); education plays an important role for both income generation and consumption behavior. We also include a variable for household size (measured in terms of AE). Moreover, we control for farm size in terms of area owned, which is a proxy for agricultural asset ownership more generally. Farm income is not included in the model, as this is directly influenced by Bt adoption. However, off-farm income, measured in US$ per year, is controlled for. We also include state dummies for Karnataka, Andhra Pradesh, and Tamil Nadu (Maharashtra is the reference state), capturing climatic and agroecological differences. Given the panel structure of the data with four survey rounds, we use year dummies for 2004, 2006, and 2008 (2002 is the reference year).

Panel data models are often estimated with a random effects estimator [31]. However, a random effects estimator can lead to biased impact estimates when there is unobserved heterogeneity between Bt adopting and non-adopting households. Such bias resulting from endogeneity of the treatment variable is referred to as selection bias in the impact assessment literature [23], [31]. Unobserved heterogeneity may potentially result from differences in household characteristics (e.g., Bt adopting farmers may have higher motivation, better management skills, or better access to information) or farm characteristics (e.g., differences in soil quality, or water access). Our panel data allow us to control for such unobserved heterogeneity. Since we surveyed the same households repeatedly over a 7-year period when Bt adoption increased, for many households we have observations with and without Bt adoption. Hence, we rely on a within household estimator, which is also called a fixed effects estimator. Differencing within households with the fixed effects estimator eliminates time-invariant unobserved factors, so that they can no longer bias the impact estimates [31]. A Hausman test is used to confirm the appropriateness of the fixed effects specification [19], [31].

We estimate an additional model using calories from more nutritious foods (i.e., pulses, fruits, vegetables, and animal products) instead of total calorie consumption as dependent variable. This additional model helps to analyze impacts of Bt cotton adoption on dietary quality. A positive coefficient for the treatment variable would indicate that Bt adoption increases the consumption of more nutritious foods, thus not only contributing to more calories but also to better dietary quality.

Results and Discussion

Descriptive statistics are shown in Table 2. The average farm household owns 5 ha of land, without a significant difference between Bt adopters and non-adopters. Around half of this area is grown with cotton. Other crops cultivated include wheat, millet, sorghum, pulses, and in some locations rice, among others. Households are relatively poor; average annual per capita consumption expenditures range between 300 and 500 US$.

Table 2. Descriptive statistics of farm households.

Variables	Adopters of Bt (N = 1065)	Non-adopters of Bt (N = 346)
Farm size (ha)	5.11 (5.85)	4.85 (5.51)
Cotton area cultivated (ha)	2.35 (2.35)	2.79 (19.67)
Area cultivated with Bt cotton (ha)	1.87*** (2.08)	0.00 (0.00)
Age of farmer (years)	45.58 (12.86)	45.94 (12.36)
Education of farmer (years)	7.58*** (4.94)	6.69 (5.03)
Per capita consumption expenditure (US$/year)	490.31*** (430.18)	311.72 (355.56)
Off-farm income (US$/year)	560.70 (1455.44)	504.27 (2299.87)
Calorie consumption per AE (kcal/day)	3329.41*** (719.38)	2629.88 (598.99)
Calories consumed from more nutritious foods per AE (kcal/day)[a]	703.89*** (374.90)	638.89 (345.41)
Household size (AE)	5.01 (2.42)	5.14 (2.24)
Food insecure households (%)[b]	7.93***	19.94

Mean values are shown with standard deviations in parentheses. N: Number of observations. AE: adult equivalent.
***Mean values between adopters and non-adopters of Bt are statistically significant at the 1% level.
[a]More nutritious foods include pulses, fruits, vegetables, and all animal products.
[b]Consumption of less than 2300 kcal per AE and day.
doi:10.1371/journal.pone.0064879.t002

doi:10.1371/journal.pone.0064879.t002

Bt adopting households consume significantly more calories than non-adopting households, and a smaller proportion of them is food insecure (Figure 1, Table 2). This suggests that the cash income gains through Bt adoption may have improved food security among cotton-producing households. Yet, this simple comparison does not yet prove a causal relationship.

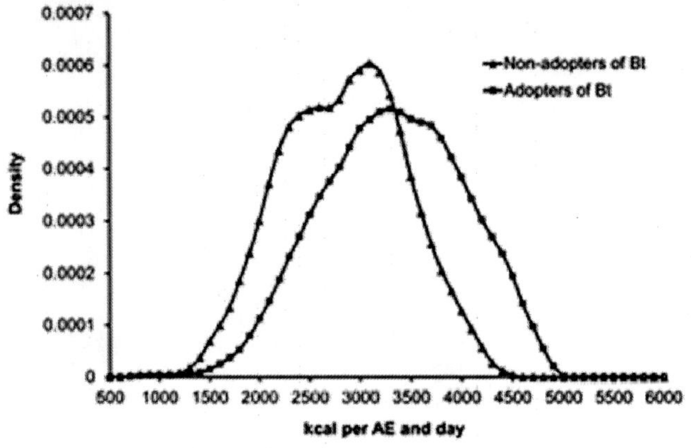

Figure 1. Density functions of household calorie consumption for adopters and non-adopters of Bt cotton. Functions were estimated non-parametrically using the Epanechnikov kernel with 1085 and 346 observations for adopting and non-adopting households, respectively. AE: adult equivalent.

doi:10.1371/journal.pone.0064879.g001

Impact of Bt Cotton Adoption on Food Security

To further analyze the relationship between Bt adoption and calorie consumption, we use panel regression models, as explained above. The main explanatory variable of interest is the Bt cotton area of a farm household, for which descriptive statistics are shown in Table 3. The average Bt area among technology adopters in the sample is close to 2 ha, which is equivalent to 85% of the total cotton area of these farms. A breakdown by survey year shows that the average Bt area increased from less than 1.0 ha in 2002 to 2.4 ha in 2008. Hence, not only the number of Bt adopters but also the Bt area per adopting household increased considerably over time.

Table 3. Bt cotton area among adopting households.

	2002	2004	2006	2008	Total
Mean Bt area (ha)	0.94	1.64	2.15	2.37	1.97
Standard deviation	1.32	1.87	2.14	2.22	2.08
Number of observations	131	246	333	375	1085

doi:10.1371/journal.pone.0064879.t003

doi:10.1371/journal.pone.0064879.t003

The regression results are shown in Table 4. Each ha of Bt cotton has increased total calorie consumption by 74 kcal per AE and day. For the average adopting household, the net effect is 145 kcal per AE (Figure 2), implying a 5% increase over mean calorie consumption in non-adopting households. Most of the calories consumed in rural India stem from cereals that are rich in carbohydrates but less nutritious in terms of protein and micronutrients. Yet the results show that Bt adoption has significantly increased the consumption of calories from more nutritious foods, thus also contributing to improved dietary quality.

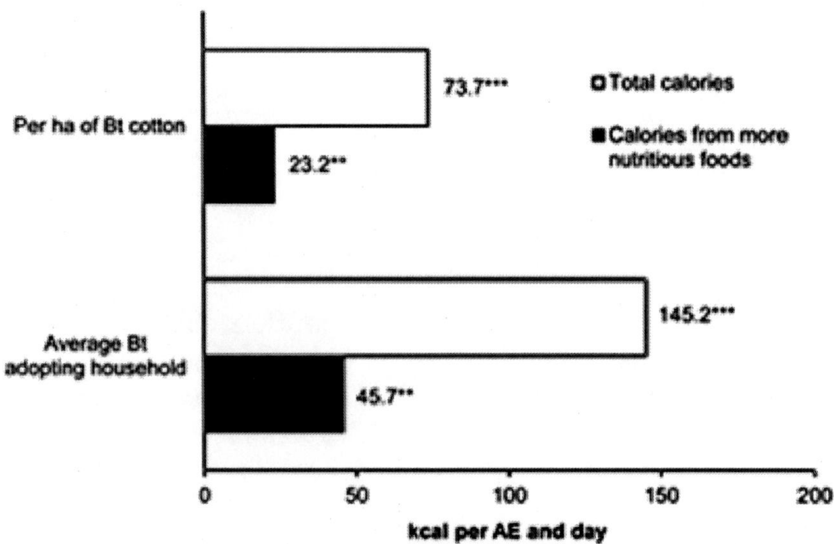

Figure 2. Net effects of Bt adoption on household calorie consumption. Results based on calorie consumption regression models estimated with panel data and household fixed effects (within estimator). Full model results are shown in Table 4. Calories from more nutritious foods include pulses, fruits, vegetables, and animal products. Effects for the average adopting household take into account the number of ha of Bt cotton actually grown. **Significant at the 5% level. ***Significant at the 1% level.

doi:10.1371/journal.pone.0064879.g002

Table 4. Calorie consumption models.

	Model (1)	Model (2)	Model (3)
Variables	Total calories (RE model)	Total calories (FE model)	Calories from more nutritious foods (FE model)
Bt area (ha)	74.28*** (16.65)	73.71*** (21.40)	23.17** (10.05)
Farm size (ha)	9.27** (4.22)	−0.64 (7.80)	1.97 (3.56)
Education of farmer (years)	9.41** (4.40)	-	-
Off-farm income (US$/year)	0.07*** (0.02)	0.05*** (0.02)	0.01* (0.007)
Household size (AE)	−62.48*** (10.71)	−89.40*** (14.63)	−29.33*** (6.89)
Karnataka (dummy)[a]	88.36 (57.97)	-	-
Andhra Pradesh (dummy)[a]	21.46 (58.30)	-	-
Tamil Nadu (dummy)[a]	212.86*** (84.56)	-	-
2004 (dummy)[b]	−34.35 (40.97)	−5.94 (51.60)	−45.25* (25.31)
2006 (dummy)[b]	13.48 (54.48)	30.09 (61.12)	−112.87*** (29.41)
2008 (dummy)[b]	−92.92 (60.51)	−74.59 (69.51)	−72.70** (30.20)
Constant	3229.31*** (90.46)	3537.08*** (76.36)	843.29*** (41.42)
Number of observations	1431	1431	1431
R^2	0.13	0.09	0.10
Hausman test (chi-square statistic)	16.62**		

The dependent variable in models (1) and (2) is the total number of kcal consumed per AE and day. The dependent variable in model (3) is the number of kcal consumed from more nutritious foods (i.e., pulses, fruits, vegetables, and animal products) per AE and day. All coefficient estimates can be interpreted as marginal effects; robust standard errors are shown in parentheses. AE: adult equivalent; RE: random effects; FE: fixed effects.
*, **, ***Significant at the 10%, 5%, and 1% level, respectively.
[a]The reference state is Maharashtra.
[b]The reference year is 2002.
doi:10.1371/journal.pone.0064879.t004

doi:10.1371/journal.pone.0064879.t004

We applied the total calorie consumption effect of Bt to the subsample of non-adopters to simulate the food security impact of adoption: if all non-adopters switched to Bt, the proportion of food insecure households would drop by 15–20% (Table 5). Most of these nutritional benefits have materialized already, as over 90% of all cotton farm households in India have adopted Bt technology by now.

Table 5. Impact of Bt adoption on food security among cotton-producing households.

	Food insecure households (%)[a]	Change in food insecurity relative to status quo (%)
Non-adopters of Bt cotton (status quo)	19.94	
If non-adopters adopted Bt on their total cotton area	15.90	−20.26
If non-adopters adopted Bt on 83% of their cotton area	16.76	−15.95

The proportion of food insecure households in the status quo refers to the subsample of 346 non-adopters. For these households, changes in calorie consumption through Bt adoption were simulated, assuming full Bt adoption (on 100% of their cotton area) and partial Bt adoption (on 83% of their cotton area, as observed in the subsample of Bt adopters). For the simulations, the net effect of Bt on total calorie consumption per ha was used (Figure 2).
[a]Consumption of less than 2300 kcal per adult equivalent and day.
doi:10.1371/journal.pone.0064879.t005

doi:10.1371/journal.pone.0064879.t005

Robustness Checks

We tested the robustness of the Bt effects by estimating calorie consumption models with alternative specifications. These additional estimates are shown in Table 6. We first look at possible changes in impact over time. In model (1), the Bt area variable is split into two periods, namely 2002–04 and 2006–08. In both periods, the Bt impact on calorie consumption was positive and significant, but the effect was bigger in 2002–04 than in 2006–08. The reason for this change is not that income effects of Bt adoption would shrink; recent research showed that the profit gains of Bt cotton in India were constant or even increased over time [23]. The change in the calorie effect per ha of Bt is rather due to the fact that the Bt area per farm increased considerably in the later period, as was shown above. Measured per farm household, the calorie consumption effect of Bt was actually very similar in 2002–04 and 2006–08.

Table 6. Robustness checks of Bt effects with different model specifications.

	Model (1)	Model (2)	Model (3)	Model (4)
Variables	Total calories	Calories from more nutritious foods	Total calories	Total calories
Bt area 2002–04 (ha)	135.25*** (28.95)	17.84 (13.24)	–	–
Bt area 2006–08 (ha)	54.67** (23.33)	24.79** (10.46)	–	–
Cumulative Bt area (ha)	–	–	17.20 (12.20)	−28.08** (13.21)
Bt area (ha)	–	–	–	105.63*** (26.82)
Number of observations	1431	1431	1431	1431
	Model (5)	Model (6)	Model (7)	Model (8)
	Total calories	Total calories	Total calories	Total calories
Bt area (ha)	73.71*** (21.40)	76.19*** (27.62)	110.81*** (27.46)	53.40* (30.99)
Bt (dummy)	–	–	–	596.84*** (70.29)
Number of observations[a]	1431	1016	852	852

All models are estimated with household fixed effects. Other explanatory variables were included in estimation, as in Table 4, but are not shown here for brevity. The dependent variable in all models is calorie consumption measured in kcal per AE and day. Coefficient estimates can be interpreted as marginal effects; robust standard errors are shown in parentheses.

*, **, ***Significant at the 10%, 5%, and 1% level, respectively.

[a]In model (6), all observations of households that had adopted Bt in 2002 were dropped. In models (7) and (8), all observations of households that had adopted Bt in all survey rounds were dropped.

doi:10.1371/journal.pone.0064879.t006

doi:10.1371/journal.pone.0064879.t006

The smaller calorie consumption effect per ha of Bt with an increasing Bt area on a farm is consistent with Engel's law, which states that the proportion of the household budget spent on food decreases as income rises [32]. Unsurprisingly, the same trend is not observed when we focus on higher value, non-staple foods. The results of model (2) in Table 6 suggest that the Bt effect on calories from more nutritious foods has been increasing over time. Hence, Bt cotton adoption leads to a lower staple calorie share, implying higher nutritional sufficiency and better dietary quality [29].

In model (3) of Table 6, we analyze whether the Bt effect is cumulative, meaning that households that have adopted Bt earlier or on larger areas benefit over-proportionally. This might be the case when profit gains from Bt adoption are reinvested, possibly entailing larger consumption benefits in subsequent periods. To test for this option, we constructed a cumulative Bt area variable, adding up the Bt area on a farm in a particular year and Bt areas on the same farm in previous survey rounds. The coefficient of this variable is insignificant; cumulative effects do not seem to be important. If we include this variable together with the standard Bt area variable, the cumulative coefficient turns negative while the actual treatment effect increases (model 4). Again, this is consistent with Engel's law, implying that larger areas with Bt lead to lower proportions of the income gains being spent on calories.

In models (6) and (7), we analyze to what extent changes in the sample affect the estimation results. For easy comparison, results from the full-sample reference model, which were discussed above, are repeated in model (5). It is sometimes observed that early adopters of a new technology benefit more than late adopters. This may be due to cumulative effects, which we already tested for. In addition, general equilibrium adjustments may contribute to differential impacts between early and late adopters [33]. In model (6), we exclude all households that had adopted Bt already in the first survey round in 2002. The change in the Bt effect is very small, so we conclude that late adopters enjoy the same nutritional benefits per ha of Bt as early adopters.

This specification in model (6) with early adopters excluded is also an additional robustness check for possible issues of endogeneity and selection bias. The fixed effects panel estimator controls for time-invariant heterogeneity between adopters and non-adopters of Bt. But it cannot control for possible time-variant differences, which might play a role if early adopters are more innovative also with respect to other opportunities not captured in our data. The similarity of the results in models (5) and (6) substantiates that the estimated Bt impacts do not suffer from selection bias. In model (7), we exclude all observations of households that had adopted Bt in all survey rounds, so that the results are purely based on within household comparisons. The treatment effect remains highly significant. It even increases in magnitude, suggesting that the full-sample result is rather a cautious, lower-bound estimate. Finally, model (8) includes a dummy for Bt adoption in addition to the Bt area variable used before. The dummy produces a large coefficient, underlining the positive food security impact of Bt adoption. But the Bt area effect remains positive and significant, too, which confirms that using a continuous treatment variable is appropriate.

Overall, the additional results with alternative specifications strengthen the findings and show that the positive impacts of Bt cotton adoption on food security in India are very robust.

CONCLUSIONS

The results of this research confirm that the income gains through Bt cotton adoption among smallholder farm households in India have positive impacts on food security and dietary quality. GM crops are not a panacea for the problems of hunger and malnutrition. Complex problems require multi-pronged solutions. But the evidence suggests that GM crops can be an important component in a broader food security strategy. So far, food security impacts are still confined to only a few concrete examples. The nutritional benefits could further increase with more GM crops and traits becoming available in the future. Appropriate policy and regulatory frameworks are required to ensure that the needs of poor farmers and consumers are taken into account and that undesirable social consequences are avoided.

ACKNOWLEDGMENTS

We thank Vijesh Krishna and two anonymous reviewers of this journal for very useful comments.

AUTHOR CONTRIBUTIONS

Analyzed the data: MQ SK. Wrote the paper: MQ SK. Conceived and designed the survey: MQ.

REFERENCES

1. FAO (2012) The State of Food Insecurity in the World (Food and Agriculture Organization of the United Nations, Rome).
2. United Nations (2012) The Millennium Development Goals Report 2012 (United Nations, New York).
3. Juma C (2011) Preventing hunger: biotechnology is key. Nature 479: 471–472. doi: 10.1038/479471a
4. Borlaug N (2007) Feeding a hungry world. Science 318: 359. doi: 10.1126/science.1151062
5. Shiva V, Barker D, Lockhart C (2011) The GMO Emperor has No Clothes (Navdanya International, New Delhi).
6. Friends of the Earth (2011) Who Benefits from GM Crops: An Industry Built on Myths (Friends of the Earth International, Amsterdam).
7. World Bank (2007) World Development Report 2008: Agriculture for Development (World Bank, Washington, DC).
8. Fedoroff NV, Battisti DS, Beachy RN, Cooper PJM, Fischhoff DA, et al. (2010) Radically rethinking agriculture for the 21st century. Science 327: 833–834. doi: 10.1126/science.1186834
9. Tester M, Langridge P (2010) Breeding technologies to increase crop production in a changing world. Science 327: 818–822. doi: 10.1126/science.1183700
10. James C (2012) Global Status of Commercialized Biotech/GM Crops: 2012, ISAAA Briefs No.44 (International Service for the Acquisition of Agri-biotech Applications, Ithaca, NY).
11. Sexton S, Zilberman D (2012) Land for food and fuel production: the role of agricultural biotechnology. In: The Intended and Unintended Effects of US Agricultural and Biotechnology Policies (eds. Zivin, G. & Perloff, J.M.), 269–288 (University of Chicago Press, Chicago).
12. Jayaraman K, Jia H (2012) GM phobia spreads in South Asia. Nature Biotechnology 30: 1017–1019. doi: 10.1038/nbt1112-1017a
13. European Commission (2010) A Decade of EU-Funded GMO Research 2001–2010 (European Commission, Brussels).
14. Paine JA, Shipton CA, Chaggar S, Howells RM, Kennedy MJ, et al. (2005) Improving the nutritional value of Golden Rice through increased pro-vitamin A content. Nature Biotechnology 23: 482–487. doi: 10.1038/nbt1112-1017a
15. Stein AJ, Sachdev HPS, Qaim M (2008) Genetic engineering for the poor: Golden Rice and public health in India. World Development 36: 144–158. doi: 10.1038/nbt1112-1017a
16. De Steur H, Gellynck X, Van Der Straeten D, Lambert W, Blancquaert

D, et al. (2012) Potential impact and cost-effectiveness of multi-biofortified rice in China. New Biotechnology 29: 432–442. doi: 10.1016/j.nbt.2011.11.012

17. Huang J, Mi J, Lin H, Wang Z, Chen R, et al. (2010) A decade of Bt cotton in Chinese fields: assessing the direct effects and indirect externalities of Bt cotton adoption in China. Science China Life Sciences 53: 981–991. doi: 10.1007/s11427-010-4036-y
18. Krishna VV, Qaim M (2012) Bt cotton and sustainability of pesticide reductions in India. Agricultural Systems 107: 47–55.
19. Kouser S, Qaim M (2011) Impact of Bt cotton on pesticide poisoning in smallholder agriculture: A panel data analysis. Ecological Economics 70: 2105–2113.
20. Qaim M (2009) The economics of genetically modified crops. Annual Review of Resource Economics 1: 665–693.
21. Ali A, Abdulai A (2010) The adoption of genetically modified cotton and poverty reduction in Pakistan. Journal of Agricultural Economics 61: 175–192. doi: 10.1111/j.1477-9552.2009.00227.x
22. Subramanian A, Qaim M (2010) The impact of Bt cotton on poor households in rural India. Journal of Development Studies 46: 295–311. doi: 10.1111/j.1477-9552.2009.00227.x
23. Kathage J, Qaim M (2012) Economic impacts and impact dynamics of Bt (*Bacillus thuringiensis*) cotton in India. Proc. Natl. Academy of Sciences USA 109: 11652–11656. doi: 10.1111/j.1477-9552.2009.00227.x
24. de Haen H, Klasen S, Qaim M (2011) What do we really know? Metrics for food insecurity and undernutrition. Food Policy 36: 760–769. doi: 10.1111/j.1477-9552.2009.00227.x
25. Gopalan C, Rama Sastri BV, Balasubramanian SC (2004) Nutritive Value of Indian Foods (Indian Council of Medical Research, National Institute of Nutrition, Hyderabad).
26. National Sample Survey Organization (2007) Nutritional Intake in India 2004–05, Report No.513 (Government of India, New Delhi).
27. Rao CHH (2005) Agriculture, Food Security, Poverty and Environment – Essays on Post-Reform India (Oxford University Press, New Delhi).
28. FAO (2001) Human Energy Requirement, Food and Nutrition Technical Report 1 (Food and Agriculture Organization of the United Nations, Rome).
29. Jensen RT, Miller NH (2010) A Revealed Preference Approach to Measuring Hunger and Undernutrition. NBER Working Paper No.16555 (National Bureau of Economic Research, Cambridge, MA).
30. Bouis HE (1994) The effect of income on demand for food in poor

countries: are our food consumption databases giving us reliable estimates? Journal of Development Economics 44: 199–226. doi: 10.1016/0304-3878(94)00012-3

31. Cameron AC, Trivedi PK (2005) Microeconometrics: Methods and Applications (Cambridge University Press, Cambridge).
32. Leathers HD, Foster P (2004) The World Food Problem: Tackling the Causes of Undernutrition in the Third World (Lynne Rienner Publishers, Boulder, CO).
33. De Janvry A, Sadoulet E (2002) World poverty and the role of agricultural technology: direct and indirect effects. Journal of Development Studies 38(4): 1–26. doi: 10.1080/00220380412331322401

Chapter 7

THE RECENT RECOMBINANT EVOLUTION OF A MAJOR CROP PATHOGEN, POTATO VIRUS Y

Johan Christiaan Visser, Dirk Uwe Bellstedt, Michael David Pirie

Department of Biochemistry, The University of Stellenbosch, Stellenbosch, South Africa

ABSTRACT

Potato virus Y (PVY) is a major agricultural disease that reduces crop yields worldwide. Different strains of PVY are associated with differing degrees of pathogenicity, of which the most common and economically important are known to be recombinant. We need to know the evolutionary origins of pathogens to prevent further escalations of diseases, but putatively reticulate genealogies are challenging to reconstruct with standard phylogenetic approaches. Currently available phylogenetic hypotheses for PVY are either limited to non-recombinant strains, represent only parts of the genome, and/or incorrectly assume a strictly bifurcating phylogenetic tree. Despite attempts to date potyviruses in general, no attempt has been made to date the origins of pathogenic PVY. We test whether diversification of the major strains of PVY and recombination between them occurred within the time frame of the domestication and modern cultivation of potatoes. In so doing, we demonstrate a novel extension of a phylogenetic approach for reconstructing reticulate evolutionary scenarios. We infer a well resolved phylogeny of 44 whole genome sequences of PVY viruses, representative of all known strains, using recombination detection and phylogenetic inference techniques. Using Bayesian molecular dating we show that the parental strains of PVY diverged around the time potatoes were first introduced to Europe, that recombination between them only occurred in the last century, and that the multiple recombination events that led to highly pathogenic PVY^{NTN}occurred within the last 50 years. Disease causing agents are often transported across the globe by humans, with disastrous effects for us, our livestock and crops. Our analytical approach is particularly pertinent for the often small recombinant genomes involved (e.g. HIV/influenza A). In the case of PVY, increased transport of diseased material is likely to blame for uniting the parents of recombinant pathogenic strains: this process needs to be minimised to prevent further such occurrences.

INTRODUCTION

Potato virus Y (PVY) afflicts potato producers worldwide [1]–[3], causing loss of yield ranging from 10% to complete crop failure. The extent of yield reduction depends on a range of factors (including the viral load, the time of infection, temperature during growth and tuber storage and the cultivar of potato that is infected [4], [5]), but the strain of PVY involved is particularly important: some are considerably more pathogenic than others [5], [6].

Whilst potatoes have been in cultivation outside the New World since the mid 16th Century, PVY was first discovered and has developed into a major crop disease only within the last 80 years. All PVY infections reduce yield, but under warmer growing conditions (such as in the potato growing regions of southern Europe and South Africa) the most detrimental strains can entirely compromise the economic viability of a crop by inducing Potato Tuber Necrotic Ringspot Disease (PTNRD). In this respect, the earliest known strains were relatively innocuous, with symptoms largely restricted to mosaic patterns or stipple streaks on leaves (PVY^{C} [7]; PVY^{O}[8]), and/or venal leaf necrosis and only rarely PTNRD (PVY^{N} [9], [10]).

More recently, genetic recombinants between PVY^{O} and PVY^{N} that induce PTNRD much more frequently have been identified. Recombination is prevalent in viruses [11]–[15] and its impact on the virulence of disease may be considerable. Foremost amongst the recently identified strains are PVY^{NTN} (N-tuber necrotic) [16] and $PVY^{N\text{-}W}$ (N-Wilga) [17], described in 1984 and 1991 respectively [18]. Both PVY^{NTN} and $PVY^{N\text{-}W}$ have spread rapidly, causing severe reductions in yields worldwide [19], [20]. In order to both limit the impact of existing strains on their hosts and, if possible, avoid creating the conditions that drive further escalation of pathogenicity, we need to understand the circumstances under which pathogenic recombinant virus strains such as these evolve.

However, evolutionary scenarios involving recombination are challenging to reconstruct. Individual 'gene trees' (phylogenies of non-recombinant regions of genomes) deviate from one another and from the underlying 'species tree' (representing the historical sequence of speciation events) due to differing underlying processes that are notoriously difficult to discern. Besides the various potential sources of analytical error (such as incorrect assessment of homology; model misspecification etc.), these include biologically meaningful processes such as reticulation (recombination between the branches of the species tree; i.e. between different species) and coalescent stochasticity (resulting from recombination within those branches, i.e. between individuals of the same species). It is important to distinguish reticulation from coalescent stochasticity in order to correctly infer species trees under the current methods that assume exclusively the latter process (e.g. [21]). The problem is compounded in viruses because despite generally high evolutionary rates [22] that might favour the availability of the necessary informative sequence variation, their diminutive genomes (e.g. PVY, 9.7 kb in length; HIV, 9.8 kb [11]; influenza A, 16.6 kb and polio, 7.4 kb [22]) represent a limited total source of data. Currently available phylogenetic hypotheses for PVY are either restricted to non-recombinant strains [23], [24] (i.e. excluding the most pertinent pathogenic ones), represent only parts of the PVY genome [25], [26], and/or incorrectly assume a strictly bifurcating phylogenetic tree [25], [27]–[29]. Despite attempts to date potyviruses in general [30], no attempt has been made to date the origins of pathogenic PVY.

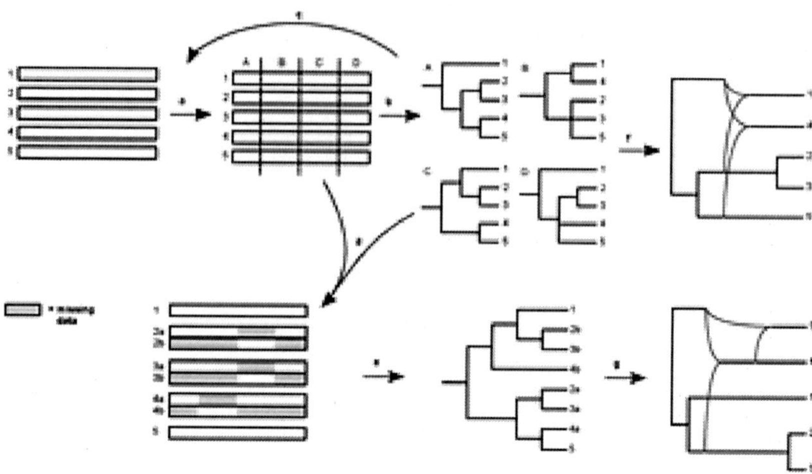

Figure 1. Summary of the analytical approach.

The aligned sequence matrix was analysed with recombination detection software (a); the resulting breakpoints were tested with standard phylogenetic analyses of the putatively non-recombinant genome regions (b), resulting in a sequence of 'gene trees' (labelled A–D), and the process repeated (c), until all topological differences between gene trees could be explained by specific recombination events and vice versa. A supermatrix was then constructed following the taxon duplication approach (d; see Fig. 2); and analysed under standard phylogenetic and (relaxed) clock models (e). Phylogenetic networks were summarised from both the separate gene trees (f) and multi-labelled 'genome tree' (g).

doi:10.1371/journal.pone.0050631.g001

In this study we reconstruct and date the phylogeny of PVY by means of a phylogenetic approach to analysing DNA sequence data in the presence of reticulation [31], [32] that we extend to address multiple recombination events between whole genomes. Unlike existing approaches, ours neither assumes a bifurcating species tree nor assumes prior knowledge of processes underlying deviations between individual gene trees. We use the resulting robust, time calibrated phylogeny to place patterns of divergence and recombination in PVY in the historical context of human cultivation of potatoes. In particular, we test whether diversification of the major strains of PVY and recombination between them occurred within the time frame of potato domestication and/or modern cultivation.

MATERIALS AND METHODS

Sampling

We sampled PVY isolates from Africa, Asia, Europe, and both North and South America, covering all known recombinant strains for which whole genome sequences were available (Table S1). Fifteen new genome sequences were generated following direct amplification RT-PCR protocols described in [33]–[35], and 29 further sequences [24], [26], [28], [36]–[46] were obtained from GenBank. Outgroups were Pepper Mottle Virus (PMV) and two isolates of Sunflower Chlorotic Mottle Virus (SCMV); the latter more closely related to PVY than those included in previous analyses (cf. [24]). Sequences were aligned using BioEdit 7.0.5.2 [47]. Short regions of uncertain homology between outgroup and ingroup sequences were treated as insertions and excluded from analyses (Dataset S1).

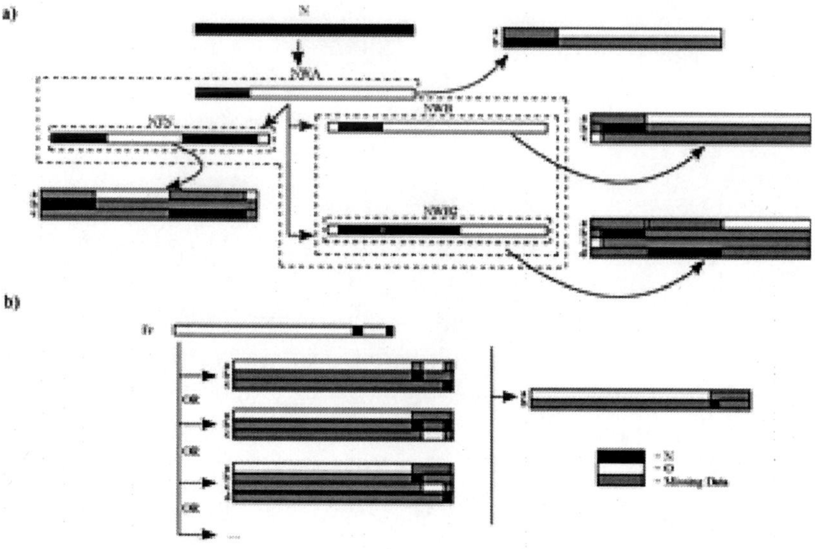

Figure 2. The taxon duplication approach and multiple recombinants.

Recombinant genomes encode a mosaic of differing phylogenetic relationships. Depending on the pattern and sequence of recombination events, separate regions of a given genome may share a common history of inheritance whilst those immediately adjacent are more distantly related. Under the 'taxon duplication' approach, these distinct 'phylogenetic signals' are segregated into separate taxa in the data matrix. Precisely which genome regions should be combined and which should be analysed independently can be inferred from the logical sequence of recombination events. a) In the case of PVY^{NW} and PVY^{NTN}, single, double and triple recombinants are apparent from shared derived recombination patterns (indicated here by black and white bars) and confirmed by exclusive ancestry (monophyly; indicated here by dotted boxes) of the pertinent genome regions. These are treated as two, three and four taxa respectively, as indicated, with the rest of the alignment re-coded as missing data. b) In the case of isolate Fr, lower recombinants are not known and phylogenetic signal is not sufficiently strong to discern congruence from conflict across the genome, thus the data could logically be combined in a number of different ways. In this case, the shorter of the non-contiguous regions are excluded from further analyses.

doi:10.1371/journal.pone.0050631.g002

Recombination Detection and Matrix Construction

Our analytical approach is illustrated in Fig. 1. We used multiple recombination detection methods as implemented in RDP3 [48] and SimPlot [12] to identify breakpoints followed by testing those breakpoints using phylogenetic analyses under parsimony and ML (as below) of non-recombinant regions to confirm the changing phylogenetic signal observed when progressing from the 5′ to the 3′ end of the linear PVY genome. Using RDP3, five methods were applied: RDP [48], [49], GENECONV [50], [51], MaxChi [52], [53], BootScan [54], [55] and SiScan [56]. Sequences were treated as linear. The threshold P-Value was set at 0.05, using Bonferroni correction. Following the RDP3 manual this should give few false positives but will still allow detection of most recombination events.

The SEQGEN parametric simulations and phylogenetic evidence options were selected. For method-specific settings we followed the RDP3 manual. The results of the subsequent phylogenetic analyses of putatively non-recombinant regions were assessed for topological conflict subject to bootstrap support (BS) ≥70% under both parsimony and ML. The process was then repeated with breakpoints that did correspond to such conflict tentatively assumed to be correct until all topological differences between the 'gene trees' could be explained by specific recombination events and vice versa. The phylogenetic analyses represent a conservative test of the (not necessarily unanimous) results of the recombination detection methods. It will tend to reject recombination both when it has been incorrectly inferred and where it is real but involves little sequence variation (generally corresponding to very short regions and/or very recent events). We regard the latter as effectively impossible to address using phylogenetic approaches and assume that it will be of low impact on the subsequent analyses.

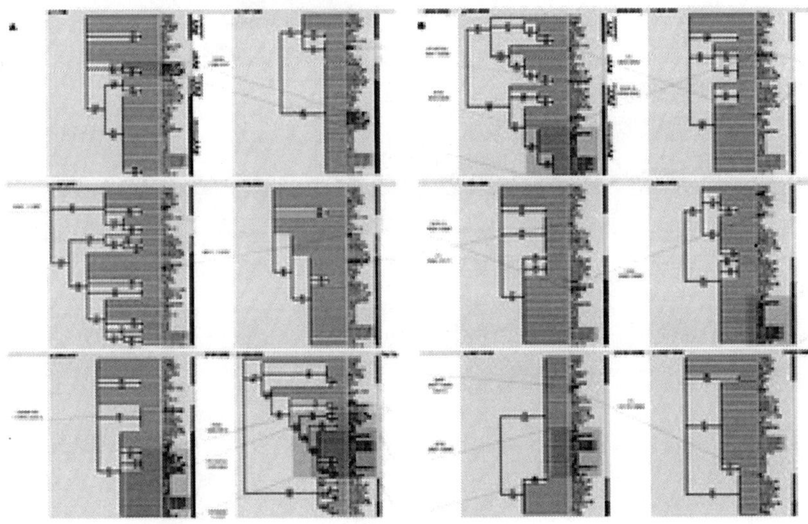

Figure 3. 'Gene' trees.

Twelve 70% BS consensus trees summarised from parsimony and maximum likelihood phylogenetic analyses of individual non-recombinant regions (presented on two pages; A and B). Differences in topologies with respect to the positions of recombinant taxa are highlighted and the corresponding recombination events indicated. The trees are presented with the major groupings (PVY$^{C/NONPOT}$, PVY^{O} and PVY$^{N\text{-North America}}$, and PVY$^{N\text{-Europe}}$), between which the recombinants switch, ordered consistently from top to bottom; these are also indicated by red, yellow, light blue and dark blue bars respectively. Where resolution of particular trees is too limited to retrieve these clades their membership - give or take recombinants - is assumed to be consistent with previous or subsequent trees.

doi:10.1371/journal.pone.0050631.g003

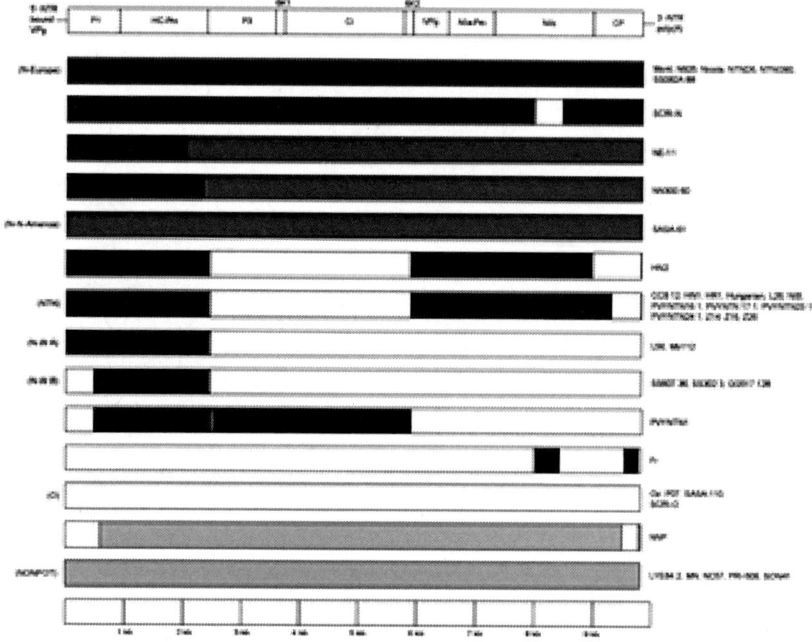

Figure 4. Recombination map of PVY genomes.

Patterns of recombination between PVY strains PVY^{O} (white), PVY$^{N\text{-North America}}$ (dark grey), PVY$^{N\text{-Europe}}$ (black) and PVY^{NONPOT} (light grey) are illustrated.

doi:10.1371/journal.pone.0050631.g004

A supermatrix was subsequently constructed in which

recombinant sequences were split into multiple taxa using the 'taxon duplication' approach [31], [32]; equivalent to the 'compatibility matrix' output format implemented in RDP3 [48] (Dataset S2). Following this approach, taxa that exhibit conflicting phylogenetic positions according to different gene regions are duplicated in the matrix and the (different) conflicting gene regions re-coded as missing data for each duplicate. The resulting supermatrix can be analyzed using standard phylogenetic techniques to produce a single 'multi-labelled' tree in which conflicting taxa - e.g. putative hybrids or recombinants - are represented more than once. This approach has previously been applied to phylogenetic analyses of conflicting gene trees in various groups of flowering plants [57]-[59], including an extension in a coalescence framework ([60] under the assumption that reticulation could be discerned from coalescent stochasticity). To our knowledge, the approach has not previously been applied to multiple recombinants or whole genomes of viruses. In order to determine which non-contiguous genome regions should be combined as single taxa in the supermatrix we first identified homologous recombination patterns, on the basis of common breakpoints and/or (given the possibility for nested recombinants) monophyly in gene trees; and then identified shared phylogenetic signal across non-contiguous genome regions (i.e. those interrupted by recombinant regions) on the basis of gene tree topological congruence and the logical sequence of homologous recombination events (Fig. 2). Where the evidence for combining non-contiguous genome regions was equivocal, we excluded from the analyses the shorter regions from the taxa in question, recoding them as unknown in the matrix (Fig. 2).

Phylogenetic Analyses

Phylogenetic analyses were performed under parsimony using PAUP* 4.0b10 [61] and under likelihood using RAxML [62]. Under parsimony, the following heuristic search options were employed: 500 random addition sequences (RAS) with tree bisection and reconnection (TBR) branch swapping saving a maximum of 25 trees of minimal length in each replicate. Clade support was estimated using 10,000 replicates of non-parametric bootstrapping each comprising a single RAS and TBR, saving a single tree in each replicate. RAxML analyses were performed using the CIPRES Science Gateway (http://www.phylo.org/portal2/) [63], [64], assuming a gamma model of rate heterogeneity. The supermatrix was analysed as above both with and without monophyly constraints on eight clades descendent from specific recombination events. This topological constraint represents the strong phylogenetic evidence of genome-scale processes that is further demonstrated by the monophyly of the clades in the separate analyses of non-recombinant gene regions. It may be important in preventing arbitrary groupings of closely related recombinant (duplicated) taxa that in the supermatrix do not share overlapping sequences. Under the tentative assumption of a reticulation scenario, phylogenetic networks were summarised using SplitsTree 4.12 [65] and Dendroscope 3 [66], 1) from the trees resulting from individual analyses of non-recombining regions and 2) from the single multi-labelled tree (in which recombinant taxa are represented more than once) resulting from analysis of the supermatrix. In both cases, nodes subject to <70% BS were first collapsed to form polytomies. Using Splitstree, consensus splits of trees were computed using the Consensus Network method [67] and splits were transformed into a reticulate network using the RECOMB2007 method [68]; cluster-based rooted networks were computed using Dendroscope.

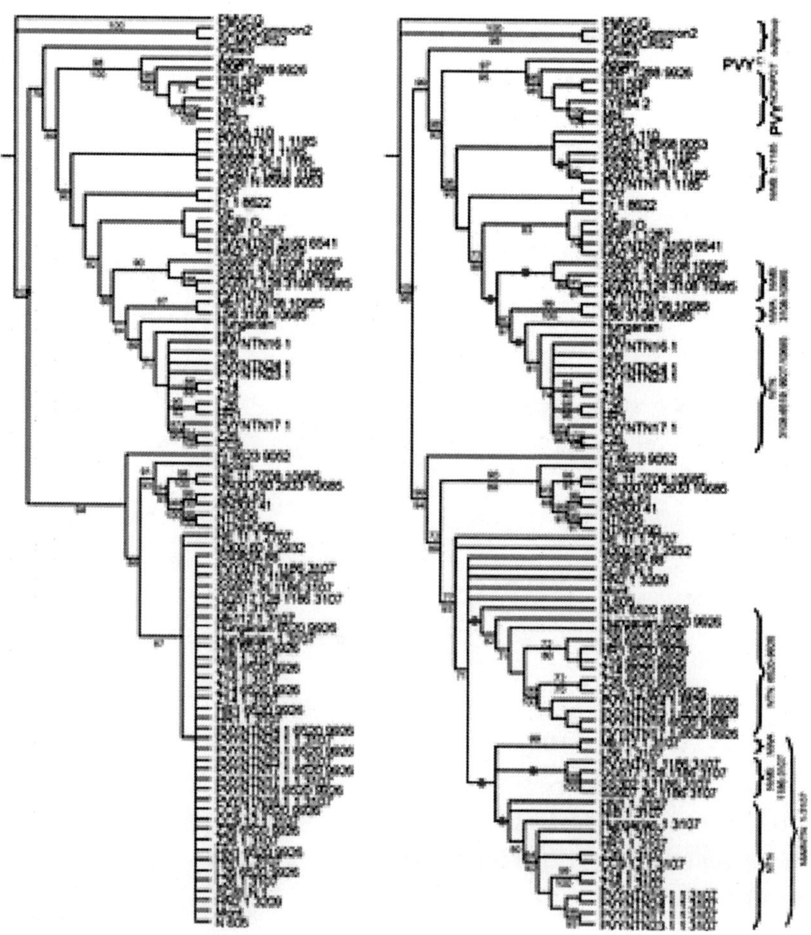

Figure 5. Supermatrix analysis.

Parsimony strict consensus trees with bootstrap support above (parsimony) and below (ML) the branches are presented; a) without; and b) with the backbone monophyly constraint. Constrained nodes are indicated by red dots on the corresponding branches.

doi:10.1371/journal.pone.0050631.g005

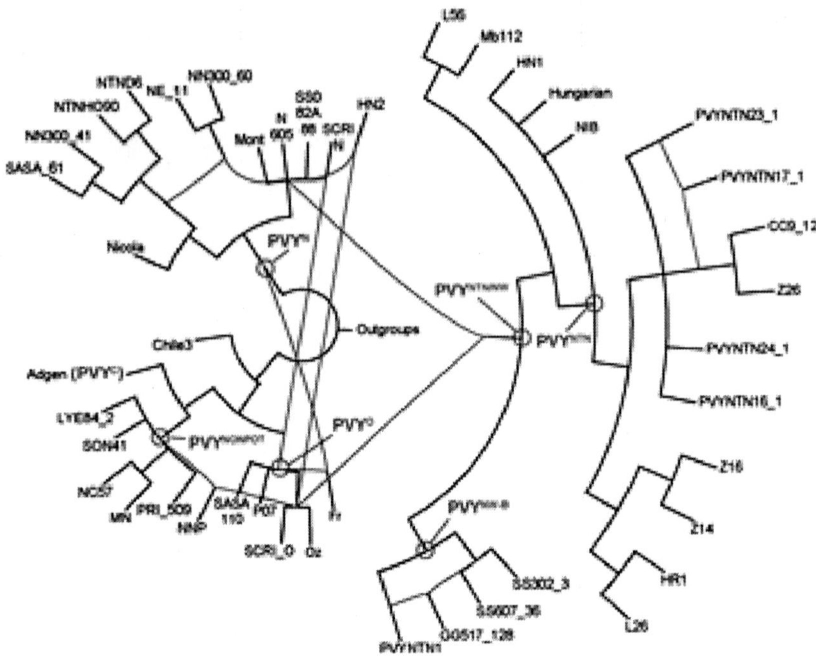

Figure 6. A reticulate phylogenetic hypothesis for PVY genomes.

A rooted phylogenetic network, assuming that recombination (blue branches) represents reticulation events. The nodes representing the most recent common ancestors of PVY strains and recombinant clades are indicated with red circles and labels.

doi:10.1371/journal.pone.0050631.g006

Molecular Dating

Path-o-gen 1.3 [21] was used to investigate the 'clocklikeness' of the PVY phylogeny with an ML tree obtained using RAxML and asynchronous tip ages (as below). The supermatrix was analysed with the above topological constraint plus a monophyly constraint for the ingroup (to root the phylogeny) using BEAST 1.7.2 [21] on the CIPRES Science Gateway [64]. We applied fixed age constraints to tips to calibrate the rate of molecular evolution [69], [70]. For sequences not produced for this study this information was obtained from the authors of the original studies; where this was not possible the isolates were omitted from the analyses (Table S1). Age estimates are more precise when the range of tip age constraints spans a higher proportion of the total age of the group [70]. The total age range of the tips in these analyses spans the years 1982 to 2010 (i.e. 28 years), which is 35% of the 80 year putative timeframe for the observation of the disease in crops, but is likely to represent a much smaller proportion of the total age of PVY root node (which is unknown). Using the taxon duplication approach, the ages of recombinant isolates contribute to calibration in multiple branches of the tree (similar to the calibration of multiple homeologues in polyploids in [71]). We applied the general time reversible (GTR) substitution model with gamma distributed rates and a proportion of invariable sites. We applied strict clock (SC) and relaxed clock models, the latter assuming lognormal (LN) and exponential distributions (EX) of rates across the phylogeny, in order to assess the sensitivity of age estimates to assumptions regarding patterns of molecular rate variation, particularly given the potential (though, to our knowledge, as yet untested) impact of missing data. Under SC two MCMC runs of 10 million generations each were performed, each sampling trees every 1,000 generations. Under LN and EX three runs of 50 or 100 million generations each were performed, sampling trees every 10,000 generations. Shorter runs excluding the recombinant sequences were performed as a joint sensitivity test for the impact of calibrations and missing data. Removing taxa from the matrix might be expected to reduce the precision of age estimates due to the directly associated loss of information regarding rate calibration (i.e. both tip ages and molecular variation). However, should the wider confidence intervals of such age estimates not contain those inferred in the presence of significant proportions of missing data this might provide evidence for some form of bias. Likelihood and topological convergence and adequate sampling of the runs were confirmed using AWTY [72] and Tracer [73]. Randomisations of tip ages as a further test of the validity of rate estimates [74] were not feasible given the lengths of runs necessary to reach convergence. We note however that datasets shown to fail such tests are generally characterised by low levels of sequence variation and/or only produce precise age estimates when constrained by informative priors, e.g. on demographic parameters or on the age of the root [74]. Neither is the case here.

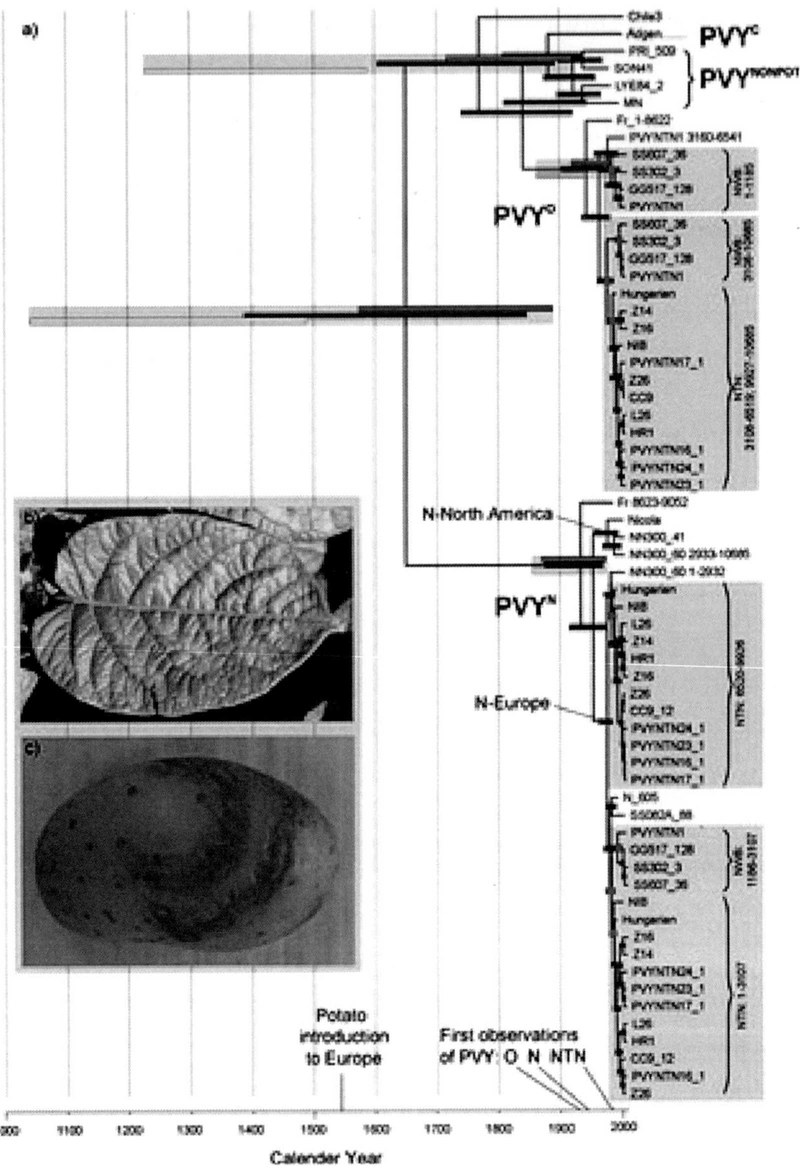

Figure 7. The recent, recombinant origins of Potato Virus Y genomes.

The maximum clade credibility tree from the BEAST LN relaxed clock analysis is shown with error bars representing the 95% HPD of node ages according to LN (blue), EX (red) and SC (yellow) models. Recombinant strains such as pathogenic PVY^{NTN} and PVY^{NW-B}are represented as multiple taxa, each representing a subset of the alignment (as indicated) with distinct phylogenetic signal. Topology constrained nodes are indicated with red dots. Inset: a potato leaf showing mosaic patterns and tuber with potato tuber necrotic ring disease resulting from PVY infection.

doi:10.1371/journal.pone.0050631.g007

Table 1. The ages of major strains of PVY.

Claderocomblnatlon	**Crown ago (years; LN/ EX/SC)**	**Stem ago (years; L4(EX/SC)**
NW/NTN 1-3107	36-21/34-20/47-27	38-22137-21/48-29
NW/NTN 3108-10654	48-21/42-20/64-35	72-27162-23179-42
NTN 1-3107	3019/2919141-25	36-21/34-20/47-27
NTN 3108-6519, 9927-10654	33-19/31-18/51-28	48-21/42-20/64-35
NTN 6520.9926	33-19/32-19/44-25	37-20/37-19/48-27
MV8 1-1185	44-10/35-8/73-24	52-14/43-10/145-69
IshY8 1186-3107	26-9/25-7/35-17	36-21/34-20/47-27
NWI3 3108-10654	19.7/23-6/19-10	48-21/42-20/64-35
FNYNTNI 31604541		52-14; 26-8; 21-6143-10; 17-5; 25-7/38-10; 79-42 35-17
NN300-60		39-10; 37-20/32-9; 37-19/98-51; 44-25
FR		106-36; 13349/93-30, 137-33)11444; 112.56

doi:10.1371/journal.pone.0050631.t001

Minimum and maiknum age estimates for the multiple recombination events leading to the origins of of PVY"w/"Th and rna,Srnum age estimates for other recombination events. The most recent bounds of the crown node age of a recombinant dade (i.e. the Most Recent Common Ancestor (mtca) of the recombinants analysed) represents the most recent possible age for that recombination event. The oldest bounds of the stem node age (Le. that of the node subtending the crown node. representing the mrca of the clade and its sister group> correspondingly represents the oldest possible age for the recombination event. 'Crown ages for recombinant dades represented multi* times in the multi-labelled tree are directly comparable, and should be expected to be the same, give or take margins of error. 'Stem ages' by contrast are dependent on taxon sampling outside the crown group and thus should be expected to vary, with some stern node ages representing a greater overestimation of the age of the recombination event than others. Therefore the oldest possible age for the recombination event can be interpreted from the older bounds of the most recent of the stem node age estimates. The minimum and maximum ages are represented in bold type. doic10.13714ournalpone.00506311002

Table 2. The ages of recombination events in PVY.

NW/NTN 1-3107	36-21/34-26/47-27
NW/NTN 3108-10654	48-21/42-20/64-35
NTN 1-3107	30-19/29-19/41-25
NTN 3108-6519, 9927-10654	33-19/31-18/51-28
NTN 6520-9926	33-19/32-19/44-25
NWB 1-1185	44-10/35-6/73-24
NWB 1186-3107	26-9/25-7/35-17
NWB 3108-10654	19-7/23-6/19-10
[illegible] 3160-6541	
[illegible]	
FR	

Minimum and maximum age estimates for the multiple recombination events leading to the origins [illegible] recombination events. The most recent bounds of the crown node age of a recombinant clade (i.e. [illegible] analysed) represents the most recent possible age for that recombination event. The oldest bounds of [illegible] node, representing the mrca of the clade and its sister group) correspondingly represents the oldest [illegible] recombinant clades represented multiple times in the multi-labelled tree are directly comparable, and [illegible] 'Stem ages' by contrast are dependent on taxon sampling outside the crown group and thus should [illegible] greater overestimation of the age of the recombination event than others. Therefore the oldest possible [illegible] older bounds of the most recent of the stem node age estimates. The minimum and maximum ages [illegible]
doi:10.1371/journal.pone.0050631.t002

doi:10.1371/journal.pone.0050631.t002

RESULTS

Recombination Breakpoints and Homologous Recombination Patterns

We analysed 44 PVY genomes with recombination detection software in order to identify recombinant isolates and locate the recombinant regions of their genomes. Just 19 of these isolates can be regarded as non-recombinant. Five are single recombinants and 20 show two or more recombination events. We located 17 breakpoints and inferred phylogenetic trees for the corresponding sequence of non-recombining regions under parsimony and Maximum Likelihood (ML) to confirm topological differences between trees (Fig. 3). We then identified homologous recombination patterns on the basis of common breakpoints and/or (given the possibility for nested recombinants) monophyly in gene trees. Both were evident for PVY^{NTN} (18 isolates) and $PVY^{N\text{-}W}$ B-type (four isolates) separately and for $PVY^{N\text{-}W}$ A- and B-types plus PVY^{NTN} together (Figs. 2 and 3). A further 10 recombination events (of 13 in total) were represented by single isolates (Table S2). Recombinant regions, the strains/isolates that exhibit them and the genome type (PVY^{NONPOT}: a strain found in various plants other than potatoes; PVY^{O} or PVY^{N}) involved are illustrated in Fig. 4.

Supermatrix Construction

In order to simultaneously infer the phylogenetic relationships of both non-recombinant and recombinant whole genomes we identified shared phylogenetic signal across non-contiguous genome regions (i.e. those interrupted by recombinant regions) on the basis of gene tree topological congruence and the logical sequence of homologous recombination events (Figs. 2and 3). On this basis, the whole genome sequences of single recombinants were subdivided between two taxa each in the matrix, double recombinants PVY^{N-W} B-type and PVY^{NTN} between three, and triple recombinant PVYNTN1 between four; each taxon representing a subset of the alignment with distinct phylogenetic signal with the rest of the alignment coded as missing data (Fig. 2). In the case of double recombinants Fr and NNP, the phylogenetic signals were not sufficiently strong to discern congruence and conflict, and ancestral-type single recombinants are unknown. Recombinant regions 9052–10654 and 10172–10654 of Fr and NNP respectively were therefore excluded from further analyses by means of recoding as missing data (Fig. 2).

Phylogenetic Inference

Phylogenetic analyses were performed on the resulting supermatrix under parsimony and ML. Of 9,723 characters included in the analyses (reduced from an alignment of 10,685 with outgroups), 5,683 were variable and 4,267 parsimony informative. In order to avoid potential loss of phylogenetic resolution between taxa with entirely non-overlapping sequences a topological constraint was designed to enforce the monophyly of each clade of homologous recombinants. This corresponded to eight nodes of a total of 89 given a 90 taxa bifurcating tree (Fig. 5). The results were entirely congruent irrespective of whether this topological constraint was applied or not, but the constrained analysis resulted in considerably higher resolution both within and between constrained clades without leading to an increase in shortest tree length (Fig. 5). Networks summarised from the individual gene trees (non-recombinant regions separately; Fig. S1 A) and the multi-labelled tree (resulting from analysis of the supermatrix; Fig. 6, Fig. S1 B) were broadly comparable, but the former considerably more complex. In both cases network structure was revealed within monophyletic PVY^{NTN} and PVY^{N-W} B clades.

Molecular Dating

Path-o-gen was used to calculate a regression of root-to-tip distances against dates of sampling. The slope of the regression (representing the rate) was 0.00237; with correlation coefficient (variation in rate) 0.1885; R squared 0.0355; and residual mean squared 0.0073; indicating deviation from a strict molecular clock. In order to infer simultaneously phylogenetic relationships and the ages of clades and recombination events we analysed the supermatrix under Bayesian inference with both strict and relaxed clock models, including only the 28 (of 44) PVY isolates for which we could obtain accurate asynchronous sampling dates. Tree samples from independent Bayesian runs of the supermatrix under each model showed consistent and stable posterior probability (PP) clade support and effective sampling sizes for model parameters >200. The maximum clade credibility tree of the LN analysis is illustrated in Fig. 7; nodes subject to ≥0.95 PP were consistent with those ≥70% BS inferred under parsimony and ML (data not shown). In general, strict clock (SC) based estimates for deeper nodes are older than those based on either lognormal (LN) or exponential (EX) models, but the discrepancy is smaller for more recent nodes. The 95% highest posterior density interval (95% HPD) for the clock/mean rate were 0.0001–0.0003 (SC); 0.0003–0.0012 (LN); and 0.0005–0.0015 (EX). The 95% HPD for the standard deviation of the LN relaxed clock was 0.6128–1.0086, which as it does not include zero further indicates the rejection of the strict molecular clock. The 95% HPD for the crown node of PVY dates to 619-161 (LN)/436-123 (EX)/970-525 (SC) years ago. Crown nodes ages of PVY^{O}, PVY^{N} and PVY^{NONPOT} are similar according to the different methods, falling between around 150 and 30 years (although PVY^{NONPOT} is older under the SC model: 365-201 years; Table 1). Excluding the recombinant isolates (leaving just 13 of 66 taxa in the supermatrix) resulted in much broader but overlapping age ranges, e.g. for the PVY crown node: 4,738-84 (LN); 3,498-58 (EX); and 133,080-801 (SC). Hence the confidence intervals extended considerably further back in time (there was no prior constraint for the age of the root node in any of the analyses), but with the exception of the SC model also increased towards the present (which is by definition constrained by the ages assigned to the tips). In the relaxed clock results based on full taxon sampling with missing data there was thus no obvious bias towards either older or more recent age estimates. The range of crown and stem node age estimates for recombinant clades (i.e. $PVY^{N\text{-}W}$/PVY^{NTN}, PVY^{NTN} and $PVY^{N\text{-}W}$ B) place the corresponding recombination events between 48 and 20; 47 and 19; and 47 and 6 years ago, respectively (Table 2). Recombination events represented by single isolates have stem ages in years as follows: PVYNTN1: 35-17; NN300_60: 44-32; and Fr: 112-93.

DISCUSSION

The most serious consequences of PVY infection are yield reduction and PTNRD. Specific genes are currently under investigation for their potential to cause pathologies [1], [5], [6], [16]but the symptoms are generally worse in the recombinant strains PVY^{NTN} and PVY^{NW} and recombination has also been directly implicated as a cause of pathogenicity [36]. Our results show that recombination is widespread amongst both highly pathogenic and less pathogenic PVY strains. We recovered essentially the same breakpoints identified for individual isolates in previous work (suggesting that our recombination detection approach was not overly conservative), as well as identifying a novel recombination pattern in NN300_60, a South African isolate similar to the previously described NE-11 [28]. The resulting phylogeny shows the major groupings of PVY strains [24], [26]–[29] and qualifies the phylogenetic affinities of the isolate Chile3 (apparently not the sister group to PVY, contra [24]), whilst simultaneously identifying the multiple phylogenetic relationships of the recombinants.

Our results confirm the single origins of recombinant PVY^{NTN} and PVY^{NW} strains, whilst at the same time providing evidence for recombination events within those strains. Both the original recombination events and some of those occurring subsequently must be associated with more or less identical breakpoints. This phenomenon has been reported for other viruses, such as begomoviruses, which have been shown to recombine at non-random breakpoints [13], due at least in part to selection [75]. The extreme consequence of such a process would be where structural genes in viruses represent 'functionally interchangeable modules with effectively independent evolution' [14]. Our results nevertheless imply tractable sequences of recombination events within an otherwise effectively tree-like underlying phylogeny. With the supermatrix approach, the phylogenetic affinities of even relatively short (non-recombinant) genome regions can be assessed within the strong phylogenetic 'scaffold' provided by the other data [31], [76]. The improved overall resolution that we obtained for the PVY phylogeny using the taxon duplication-based supermatrix compared to the separate analyses of non-recombinant genome regions is reflected in a simpler network with far fewer alternative connections between recombinants and non-recombinants (Fig. S1).

Further investigation of our approach (and of supermatrix analyses in general) should address the potential impact of missing data on age estimation. Although no obvious bias was apparent here (with the exception of the arguably inappropriate SC model), the power of the assessment was limited by the small number of non-recombinant genomes analysed. Assuming that the models implemented in BEAST do adequately reflect the uncertainty involved, the advantages of our approach are clear: by including both non-recombinant and recombinant isolates in single molecular dating analyses in this manner we were able to estimate the ages of both clades and recombination events between them, with improved precision relative to analysis of non-recombinants alone.

The potato, *Solanum tuberosum* L., originated in the New World and its centre of diversity is in the Andes. All cultivated varieties descend from a single domestication [77], with the first archaeological evidence of potato use dating to ca. 750 BC in Peru and the first likely cultivation of potatoes dating to 400 AD [78]. However, our age estimates for the most recent common ancestor (mrca) of PVY correspond more closely to the timing of the first introductions of potatoes to Europe (between 1540 and 1565 to Spain and in 1565 to Britain [79]), consistent with the recent origins inferred for various plant viruses [80]. The age estimates could indicate origin either in the New World before the European introduction, or thereafter, but given that our oldest age estimates were produced under the SC model, which appears less appropriate for this data, the latter seems more plausible. Age estimates for deeper nodes in the phylogeny are imprecise, which is inevitable given the relatively recent asynchronous tip ages used for the molecular clock calibration. However, our calibration is independent of any assumptions regarding correlations of particular phylogenetic relationships with historical isolation and outbreak events (as used by [30] in their study of potyviruses). This logical independence is crucial given our aim to test exactly these kinds of hypotheses. Despite the wide confidence intervals, even the oldest estimates for the ages of the crown nodes of PVY^O and PVY^N fall within the last c. 150 years. This places the timeframe of the radiation of PVY clades known from potato crops, as well as all known recombination events between those clades, well within that of modern potato cultivation.

In fact, most recombination events inferred here are conspicuously recent. PVY^{N-W} A-type descends from a common recombination between PVY^{O} and PVY^{N} which dates to 48-20 years ago, and PVY^{NTN} and PVY^{N-W} B-type are the results of two subsequent recombination events that we date to between 47 and 19 years ago and 47 and 6 years ago, respectively. These ages are consistent with the first observations of symptoms associated with specific strains in potato crops. PVY strains isolated from non-potato hosts are restricted to the $PVY^{NON-POT}$clade and isolate Chile3 [24] and gene flow (in the form of recombination) between these and other PVY clades appears to be rare compared to that between and within PVY^{O} and PVY^{N}. Overall, these suggest that the strains of PVY currently infecting crops have evolved as specialists of potato cultivars and not, as might have been the case, by lateral transfer from other hosts. They are consistent with recent (recombinant) origins of pathogenic strains of PVY within modern potato crops. Given the age estimates, these may have been particularly associated with increased 20th Century international trade; there is no evidence for earlier recombination between the major PVY strains.

Advances in transport have inevitably led to the increasingly rapid distribution of material infected with different strains of PVY^{O} and PVY^{N}. It is clear from the age estimates presented here, as well as the high genetic diversity of PVY strains found in individual countries such as South Africa [33], that measures to control such movement were implemented subsequent to the origin and spread of the most damaging PVY recombinant strains. Our results illustrate how recombination between both distantly and closely related strains of PVY has contributed to the origin of pathogenic strains such as PVY^{NTN}. They also provide evidence for ongoing recombination within PVY^{NTN}, as might be expected from patterns observed in other viruses[11]–[14]. This process is a cause for concern in the context of disease prevention, as it could facilitate combinations of sequence variants that increase virus fitness, for example by improving transmission. For crop plants, it is possible at least in principle to reduce the spread of diseased material by stringent testing as part of national certification schemes and monitoring of imports. Our results serve to further highlight the importance of such efforts.

CKNOWLEDGMENTS

Aelys Humphreys provided comments on a draft of the manuscript. Two anonymous reviewers and Academic Editor Darren Martin provided further constructive criticism. The authors are grateful to various researchers who provided additional information regarding their published sequences.

AUTHOR CONTRIBUTIONS

Conceived and designed the experiments: MDP DUB. Performed the experiments: JCV. Analyzed the data: MDP JCV. Contributed reagents/materials/analysis tools: DUB. Wrote the paper: MDP JCV DUB.

REFERENCES

1. McDonald JD, Singh RP (1996) Host range, symptomology, and serology of isolates of potato virus Y (PVY) that share properties with both the PVY^N and PVY^O strain groups. The American Potato Journal 73: 309–313. doi: 10.1007/bf02855210
2. Ward CW, Shukla DD (1991) Taxonomy of potyviruses: current problems and some solutions. Intervirology 32: 269–296. doi: 10.1007/bf02855210
3. Valkonen JP (2007) Viruses: economical losses and biotechnological potential. In: Vreugdenhil D, Bradshaw J, Gebhardt C, Govers F, Taylor M et al.., editors. Potato biology and biotechnology: Advances and perspectives. Amsterdam: Elsevier. 619–641.
4. Warren M, Krüger K, Schoeman AS (2005) Potato virus Y (PVY) and potato leaf roll virus (PLRV): Literature review for potatoes South Africa. Department of Zoology and Entomology, Faculty of Natural and Agricultural Sciences, University of Pretoria.
5. Le Romancer M, Kerlan C, Nedellec M (1994) Biological characterisation of various geographical isolates of potato virus Y inducing superficial necrosis on potato tubers. Plant Pathology 43: 138–144. doi: 10.1111/j.1365-3059.1994.tb00563.x
6. Boonham N, Walsh K, Hims M, Preston S, North J, et al. (2002) Biological and sequence comparisons of Potato virus Y isolates associated with potato tuber necrotic ringspot disease. Plant Pathology 51: 117–126.

doi: 10.1111/j.1365-3059.1994.tb00563.x

7. Salaman RN (1930) Virus diseases of potato: Streak. Nature 126: 241. doi: 10.1111/j.1365-3059.1994.tb00563.x
8. Smith KM (1931) Composite nature of certain potato viruses of the mosaic group. Nature 127: 702. doi: 10.1111/j.1365-3059.1994.tb00563.x
9. Orlando A, Silberschmidt K (1945) Estudos sôbre a transmissao da doenca de Solanaceas "Necroses das Nervuras" . por afïdios, e algumas relacoes entre esse virus e o seu principle inseto vector. Arquivos do Instituto Biológico 6: 133–152. doi: 10.1111/j.1365-3059.1994.tb00563.x
10. Crosslin JM, Hamm PB, Shiel PJ, Hane DC, Brown CR, et al. (2005) Serological and molecular detection of tobacco veinal necrosis isolates of Potato virus Y (PVY^N) from potatoes grown in the western United States. American Journal of Potato Research 82: 263–269. doi: 10.1007/bf02871955
11. Castro-Nallar E, Pérez-Losada M, Burton GF, Crandall KA (2012) The evolution of HIV: Inferences using phylogenetics. Molecular Phylogenetics and Evolution 62: 777–792. doi: 10.1016/j.ympev.2011.11.019
12. Lole KS, Bollinger RC, Paranjape RS, Gadkari D, Kulkarni SS, et al. (1999) Full-Length Human Immunodeficiency Virus Type 1 Genomes from Subtype C-Infected Seroconverters in India, with Evidence of Intersubtype Recombination. J Virol 73: 152–160. doi: 10.1007/bf02871955
13. Lefeuvre P, Martin DP, Hoareau M, Naze F, Delatte H, et al. (2007) Begomovirus 'melting pot' in the south-west Indian Ocean islands: molecular diversity and evolution through recombination. J Gen Virol 88: 3458–3468. doi: 10.1099/vir.0.83252-0
14. Heath L, van der Walt E, Varsani A, Martin DP (2006) Recombination Patterns in Aphthoviruses Mirror Those Found in Other Picornaviruses. J Virol 80: 11827–11832. doi: 10.1128/JVI.01100-06
15. Tan Z, Wada Y, Chen J, Ohshima K (2004) Inter- and intralineage recombinants are common in natural populations of Turnip mosaic virus. Journal of General Virology 85: 2683–2696. doi: 10.1099/vir.0.80124-0
16. Beczner L, Horvath J, Romhanyi I, Forster H (1984) Studies on the etiology of tuber necrotic ringspot disease in potato. Potato Research 27: 339–352. doi: 10.1007/bf02357646
17. Chrzanowska M (1991) New isolates of the necrotic strain of potato virus Y (PVY^N) found recently in Poland. Potato Research 34: 179–182. doi: 10.1007/bf02357646

18. Singh RP, Valkonen JPT, Gray SM, Boonham N, Jones RAC, et al. (2008) Discussion paper: The naming of *Potato virus Y* strains infecting potato. Archives of Virology 153: 1–13. doi: 10.1007/s00705-007-1059-1

19. Rahman MS, Akanda AM (2009) Performance of seed potato produced from sprout cutting, stem cutting and conventional tuber against PVY and PLRV. Bangladesh Journal of Agricultural Research 34: 609–622. doi: 10.1007/bf02357646

20. Blanchard A, Rolland M, Delaunay A, Jacqout E (2008) An international organization to improve knowledge on Potato virus Y. In: Tennant P, Benkeblia N, Blanchard A, Rolland M, Delaunay A et al., editors. Potato II Fruit, Vegetable and Cereal Science and Biotechnology 3: 6–9.

21. Drummond AJ, Rambaut A (2007) BEAST: Bayesian evolutionary analysis by sampling trees. BMC Evolutionary Biology 7: 214. doi: 10.1186/1471-2148-7-214

22. Drake JW (1993) Rates of spontaneous mutation among RNA viruses. Proceedings of the National Academy of Sciences 90: 4171–4175.

23. Cuevas JM, Delaunay A, Visser JC, Bellstedt DU, Jacquot E, et al.. (2012) Phylogeography and molecular evolution of Potato virus Y. PLoS ONE.

24. Moury B (2010) A new lineage sheds light on the evolutionary history of *Potato virus Y*. Molecular Plant Pathology. 11: 161–168.

25. Galvino-Costa SBF, Dos Reis Figueira A, Camargos VV, Geraldino PS, Hu XJ, et al.. (2011) A novel type of Potato virus Y recombinant genome, determined for the genetic strain PVYE. Plant Pathology: 1–11.

26. Ogawa T, Tomitaka Y, Nakagawa A, Ohshima K (2007) Genetic structure of a population of Potato virus Y inducing potato tuber necrotic ringspot disease in Japan; comparison with North American and European populations. Virus Research 131: 199–212. doi: 10.1016/j.virusres.2007.09.010

27. Hu X, Meacham T, Ewing L, Gray SM, Karasev AV (2009) A novel recombinant strain of Potato virus Y suggests a new viral genetic determinant of vein necrosis in tobacco. Virus Research 143: 68–76. doi: 10.1016/j.virusres.2009.03.008

28. Lorenzen JH, Nolte P, Martin D, Pasche JS, Gudmestad NC (2008) NE-11 represents a new strain variant class of Potato virus Y. Archives of Virology. 153: 517–525. doi: 10.1136/gut.2008.169052

29. Karasev AV, Hu X, Brown CJ, Kerlan C, Nikolaeva OV, et al. (2011) Genetic diversity of the Ordinary Strain of Potato virus Y (PVY) and origin of recombinant PVY strains. Phytopathology 101: 778–785. doi: 10.1094/PHYTO-10-10-0284

30. Gibbs AJ, Ohshima K, Phillips MJ, Gibbs MJ (2008) The Prehistory of Potyviruses: Their Initial Radiation Was during the Dawn of Agriculture. PLoS ONE 3: e2523. doi: 10.1371/journal.pone.0002523

31. Pirie MD, Humphreys AM, Galley C, Barker NP, Verboom GA, et al. (2008) A novel supermatrix approach improves resolution of phylogenetic relationships in a comprehensive sample of danthonioid grasses. Molecular Phylogenetics and Evolution 48: 1106–1119. doi: 10.1016/j.ympev.2008.05.030

32. Pirie MD, Humphreys AM, Barker NP, Linder HP (2009) Reticulation, data combination, and inferring evolutionary history: an example from Danthonioideae (Poaceae). Systematic Biology 58: 612–628. doi: 10.1093/sysbio/syp068

33. Visser JC, Bellstedt DU (2009) An assessment of molecular variability and recombination patterns in South African isolates of Potato virus Y. Archives of Virology. 154: 1891–1900. doi: 10.1007/s00705-009-0525-3

34. Lorenzen JH, Piche LM, Gudmestad NC, Meacham T, Shiel P (2006) A Multiplex PCR assay to characterize Potato virus Y Isolates and identify strain mixtures. Plant Disease 90: 935–940. doi: 10.1007/s00705-009-0525-3

35. Bellstedt DU, Pirie MD, Visser JC, de Villiers MJ, Gehrke B (2010) A rapid and inexpensive method for the direct PCR amplification of DNA from plants. Am J Bot 97: e65–68. doi: 10.3732/ajb.1000181

36. Barker H, McGeachy KD, Toplak N, Gruden K, Žel J, et al. (2009) Comparison of genome sequence of PVY isolates with biological properties. American Journal of Potato Research 86: 227–238. doi: 10.1007/s12230-009-9076-0

37. Fanigliulo A, Comes S, Pacella R, Harrach B, Martin DP, et al. (2005) Characterisation of Potato virus Y nnp strain inducing veinal necrosis in pepper: a naturally occurring recombinant strain of PVY. Archives of Virology 150: 709–720. doi: 10.1007/s00705-004-0449-x

38. Fellers JP, Tremblay D, Handest MF, Lommel SA (2002) The Potato virus Y M(S)N(R) NIb-replicase is the elicitor of a veinal necrosis-hypersensitive response in root knot nematode resistant tobacco. Molecular Plant Pathology 3: 145–152. doi: 10.1046/j.1364-3703.2002.00106.x

39. Hu X, He C, Xiao Y, Xiong X, Nie X (2009) Molecular characterization and detection of recombinant isolates of potato virus Y from China. Archives of Virology 154: 1303–1312. doi: 10.1007/s00705-009-0448-z

40. Jakab G, Droz EBG, Baulcombe D, Malnoe P (1997) Infectious in vivo and in vitro transcripts from a full-length cDNA clone of PVY-N605, a Swiss necrotic isolate of potato virus Y. Journal of General Virology. 78: 3141–3145.

41. Moury B, Morel C, Johansen E, Jacquemond M (2002) Evidence for diversifying selection in Potato virus Y and in the coat protein of other potyviruses. Journal of General Virology 83: 2563–2573.

42. Nie X, Singh RP, Singh M (2004) Molecular and pathological characterization of N:O isolates of the Potato virus Y from Manitoba, Canada. Canadian Journal of Plant Pathology 26: 573–583.

43. Robaglia C, Durand-Tardif M, Tronchet M, Boudazin G, Astier-Manifacier S, et al. (1989) Nucleotide sequence of Potato virus Y (Strain N) genomic RNA. Journal of General Virology 70: 935–947. doi: 10.1099/0022-1317-70-4-935

44. Schubert J, Fomitcheva V, Sztangret-Wisniewska J (2007) Differentiation of Potato virus Y strains using improved sets of diagnostic PCR-primers. Journal of Virological Methods 140.

45. Singh M, Singh RP (1996) Nucleotide sequence and genome organization of a Canadian isolate of the common strain of potato virus Y (PVYO). Canadian Journal of Plant Pathology 18: 209–214.

46. Thole V, Dalmay T, Burgyan J, Balazs E (1993) Cloning and sequencing of potato virus Y (Hungarian isolate) genomic RNA. Gene 123: 149–156. doi: 10.1016/0378-1119(93)90118-M

47. Hall TA (1999) BioEdit: a user-friendly biological sequence alignment editor and analysis program for Windows 95/98/NT; 95–98.

48. Martin DP, Lemey P, Lott M, Moulton V, Posada D, et al. (2010) RDP3: a flexible and fast computer program for analyzing recombination. Bioinformatics 26: 2462–2463. doi: 10.1093/bioinformatics/btq467

49. Martin D, Rybicki E (2000) RDP: detection of recombination amongst aligned sequences. Bioinformatics 16: 562–563. doi: 10.1093/bioinformatics/16.6.562

50. Padidam M, Sawyer S, Fauquet CM (1999) Possible emergence of new geminiviruses by frequent recombination. Virology 265: 218–225. doi: 10.1006/viro.1999.0056

51. Sawyer S (1989) Statistical tests for detecting gene conversion. Molecular Biology and Evolution 6: 526–538.

52. Maynard Smith J (1992) Analyzing the mosaic structure of genes. Journal of Molecular Evolution 34: 126–129.

53. Posada D (2001) Unveiling the molecular clock in the presence of recombination. Molecular Biology and Evolution 18: 1976–1978. doi:

10.1093/oxfordjournals.molbev.a003738

54. Martin DP, Posada D, Crandall KA, Williamson C (2005) A modified bootscan algorithm for automated identification of recombinant sequences and recombination breakpoints. AIDS Research and Human Retroviruses 21: 98–102. doi: 10.1089/aid.2005.21.98

55. Salminen MO, Carr JK, Burke DS, McCutchan FE (1995) Identification of breakpoints in intergenotypic recombinants of HIV type 1 by BOOTSCANning. AIDS Research and Human Retroviruses 11: 1423–1425. doi: 10.1089/aid.1995.11.1423

56. Gibbs MJ, Armstrong JS, Gibbs AJ (2000) Sister-Scanning: a Monte Carlo procedure for assessing signals in recombinant sequences. Bioinformatics 16: 573–582. doi: 10.1093/bioinformatics/16.7.573

57. Pelser PB, Kennedy AH, Tepe EJ, Shidler JB, Nordenstam B, et al. (2010) Patterns and causes of incongruence between plastid and nuclear Senecioneae (Asteraceae) phylogenies. American Journal of Botany 97: 856–873. doi: 10.3732/ajb.0900287

58. Antonelli A, Humphreys AM, Lee WG, Linder HP (2010) Absence of mammals and the evolution of New Zealand grasses. Proceedings of the Royal Society B: Biological Sciences 278: 675–701. doi: 10.1098/rspb.2010.1145

59. Humphreys AM, Antonelli A, Pirie MD, Linder HP (2011) Ecology and evolution of the diaspore "burial syndrome". Evolution 65: 1163–1180. doi: 10.1111/j.1558-5646.2010.01184.x

60. Blanco-Pastor JL, Vargas P, Pfeil BE (2012) Coalescent Simulations Reveal Hybridization and Incomplete Lineage Sorting in Mediterranean *Linaria*. PLoS ONE 7: e39089. doi: 10.1371/journal.pone.0039089

61. Swofford DL (2003) PAUP*: Phylogenetic Analysis Using Parsimony (*and Other Methods), version 4. Sunderland, Mass.: Sinauer Associates.

62. Stamatakis A (2006) RAxML-VI-HPC: maximum likelihood-based phylogenetic analyses with thousands of taxa and mixed models. Bioinformatics 22: 2688–2690. doi: 10.1093/bioinformatics/btl446

63. Stamatakis A, Hoover P, Rougemont J (2008) A Rapid Bootstrap Algorithm for the RAxML Web Servers. Systematic Biology 57: 758–771. doi: 10.1080/10635150802429642

64. Miller MA, Pfeiffer W, Schwartz T (2010) Creating the CIPRES Science Gateway for inference of large phylogenetic trees; New Orleans, LA. 1–8.

65. Huson DH, Bryant D (2006) Application of phylogenetic networks in

evolutionary studies Molecular Biology and Evolution. 23: 254–267. doi: 10.1093/molbev/msj030

66. Huson D, Richter D, Rausch C, Dezulian T, Franz M, et al. (2007) Dendroscope: An interactive viewer for large phylogenetic trees. BMC Bioinformatics 8: 460–460. doi: 10.1186/1471-2105-8-460

67. Holland BR, Benthin S, Lockhart PJ, Moulton V, Huber KT (2008) Using supernetworks to distinguish hybridization from lineage-sorting. BMC Evolutionary Biology 8: 202–202. doi: 10.1186/1471-2148-8-202

68. Huson DH, Klöpper TH (2007) Beyond Galled Trees - Decomposition and Computation of Galled Networks. Research in Computational Molecular Biology. In: Speed T, Huang H, Huson DH, Klöpper TH, editors. Lecture Notes in Computer Science. Berlin, Heidelberg: Springer. 211–225.

69. Drummond AJ, Ho SYW, Phillips MJ, Rambaut A (2006) Relaxed Phylogenetics and Dating with Confidence. PLoS Biology 4: e88. doi: 10.1371/journal.pbio.0040088

70. Lemey P, Rambaut A, Drummond AJ, Suchard MA (2009) Bayesian Phylogeography Finds Its Roots. PLoS Computational Biology 5: e1000520. doi: 10.1371/journal.pcbi.1000520

71. Marcussen T, Jakobsen KS, Danihelka J, Ballard HE, Blaxland K, et al. (2012) Inferring species networks from gene trees in high-polyploid North American and Hawaiian violets (Viola, Violaceae). Systematic Biology 61: 107–126. doi: 10.1093/sysbio/syr096

72. Wilgenbusch JC, Warren DL, Swofford DL (2004) AWTY: a system for graphical exploration of MCMC convergence in Bayesian phylogenetic inference. Available:http://king2.scs.fsu.edu/CEBProjects/awty/awty_start.php. Accessed 2012 Oct 29.

73. Rambaut A, Drummond AJ (2003) Tracer v. 1.5. Available:http://tree.bio.ed.ac.uk/software/tracer/. Accessed 2012 Oct 29.

74. Ho SYW, Lanfear R, Phillips MJ, Barnes I, Thomas JA, et al. (2011) Bayesian Estimation of Substitution Rates from Ancient DNA Sequences with Low Information Content. Systematic Biology 60: 366–375. doi: 10.1093/sysbio/syq099

75. Lefeuvre P, Lett JM, Varsani A, Martin DP (2009) Widely Conserved Recombination Patterns among Single-Stranded DNA Viruses. The Journal of Virology 83: 2697–2707. doi: 10.1128/jvi.02152-08

76. Wiens JJ (2006) Missing data and the design of phylogenetic analysis. Journal of Biomedical Informatics 39: 34–42. doi: 10.1016/j.jbi.2005.04.001

77. Spooner DM, McLean K, Ramsay G, Waugh R, Bryan GJ (2005) A single domestication for potato based on multilocus amplified fragment length polymorphism genotyping. Proceedings of the National Academy of Sciences of the United States of America 102: 14694–14699. doi: 10.1073/pnas.0507400102
78. Towle MA (1961) The Ethnobotany of Pre-Columbian Peru. Chicago: Aldine.
79. Brücher H (1975) Domestikation und Migration von *Solanum tuberosum* L. Genetic Resources and Crop Evolution. 23: 11–74. doi: 10.1128/jvi.02152-08
80. Gibbs AJ, Fargette D, García-Arenal F, Gibbs MJ (2010) Time - the emerging dimension of plant virus studies. Journal of General Virology 91: 13–22. doi: 10.1099/vir.0.015925-0

Chapter 8

IN SILICO IDENTIFICATION AND EXPERIMENTAL VALIDATION OF INSERTION–DELETION POLYMORPHISMS IN TOMATO GENOME

Jingjing Yang, Yuanyuan Wang, Huolin Shen and Wencai Yang*

Beijing Key Laboratory of Growth and Developmental Regulation for Protected Vegetable Crops, Department of Vegetable Science, China Agricultural University, No. 2 Yuanmingyuan Xilu, Beijing 100193, China

* To whom correspondence should be addressed. Tel. +86 10-62734136. Fax. +86 10-62733404. E-mail: yangwencai@cau.edu.

ABSTRACT

Comparative analysis of the genome sequences of *Solanum lycopersicum*variety Heinz 1706 and *S. pimpinellifolium* accession LA 1589 using MUGSY software identified 145 695 insertion–deletion (InDel) polymorphisms. A selected set of 3029 candidate InDels (≥2 bp) across the entire tomato genome were subjected to PCR validation, and 82.4% could be verified. Of 2272 polymorphic InDels between LA 1589 and Heinz 1706, 61.6, 45.2, and 31.6% were polymorphic in 8 accessions of *S. pimpinellifolium*, 4 accessions of *S. lycopersicum* var. *cerasiforme*, and 10 varieties of *S. lycopersicum*, respectively. Genetic distance was 0.216 in *S. pimpinellifolium*, 0.202 in *S. lycopersicum* var. *cerasiforme*, and 0.108 in*S. lycopersicum*. The data suggested a reduction of genetic variation fromS. *pimpinellifolium* to S. *lycopersicum* var. *cerasiforme* and S. *lycopersicum*. Cluster analysis showed that the 8 accessions of S. *pimpinellifolium* were in one group, whereas 4 accessions of S. *lycopersicum* var. *cerasiforme* and 10 varieties of S. *lycopersicum* were in the same group.

INTRODUCTION

Tomato (Solanum lycopersicum L.) is an economically important vegetable crop worldwide and a pre-eminent plant genetic analysis system. Genetic marker development for tomato has been conducted over 30 years through various approaches, including restriction fragment lengthpolymorphism (RFLP), random amplified polymorphic DNA (RAPD), amplified fragment length polymorphisms (AFLPs), simple sequence repeat (SSR), cleaved amplifiedpolymorphisms(CAPs),andconservedorthologsets(COSs). Most markers developed by these approaches are based on DNA or cDNA polymorphisms between wild species and cultivated tomato, which lead to the construction of the first generation reference linkage maps and isolation of genes of interests.1,2 However, the ability of using these markers to detect polymorphisms in cultivated tomato is limited.3 Recent efforts to develop new markers in cultivated tomato have been focus on single-nucleotide polymorphisms (SNPs) using in silico mining of expressed sequence tag database and experimental validation,4–7 amplicon sequencing of COS genes,8,9hybridization to oligonucleotide array,10 and next-generation sequencing of transcriptome or re-sequencing of genome.11–13 Owing to the abundance and wide distribution of SNPs in the whole genome and the availability of automatic large-scale genotyping platform, SNPs have widely been used in association analysis,13–15 high-density SNP map construction,7,16 as well as population structure and genetic variation analysis17–20 in cultivated tomato.

Short insertion and deletion (InDel) polymorphisms are increasingly being received attention in human because they are the second abundant form of genetic variation and can influence multiple human phenotypes including diseases.21–25 Therefore, great efforts have been put on identification, mapping, and functional analysis of InDels in the human genome.26–28 Similar work has been done in other species, such as Arabidopsis and rice.29–33 In tomato, a total of 749 966 putative InDels of 3–300 bp have been identified by comparing the genome sequences ofSolanum pimpinellifolium accession LA 1589 and S. lycopersicum variety Heinz 1706,34 and more than 80 000 putative InDels of 1–15 bp have been discovered by comparative analysis of transcriptome between wild species S. galapagense and cultivated tomato.35 However, less work on discovery of InDels in cultivated tomato has been done.

The availability of the whole genome or transcriptome sequences provides a potential to identify InDels in silico. We here developed a pipeline to identify InDels by comparative analysis of the two available genome sequences of LA 1589 and Heinz 1706. A total of 3029 candidate InDels were subjected to experimental validation by PCR amplification of genomic DNA in a collection of 22 tomato lines. The main objective of this study was to develop easy-using markers for genetic study and marker-assisted selection in cultivated tomato.

MATERIALS AND METHODS

Plant materials and DNA isolation

A panel of 22 tomato genotypes comprising of cultivated tomato (S. lycopersicum) and its wild relatives were used to validate InDel polymorphisms. These inbred lines were selected to represent a diverse collection including eight accessions of S. pimpinellifolium, five processing varieties, one greenhouse cultivar, four fresh market cultivars, and four S. lycopersicum var. cerasiforme accessions (Table 1). Nine of them were used for SNP detection in our previous study.9 The eight S. pimpinellifolium accessions were selected from the core collection or sources being used for genetic studies and were used to detect polymorphisms of candidate InDels within the species. Genomic DNA was isolated from fresh-collected young leaves of at least eight plants for each genotype using the modified CTAB method.36

Table 1. Description of plant materials

Genotype	Species	Market type	Origin	Note
LA 1269	*Solanum pimpinellifolium*	Wild	Peru	Resistance source for late blight (*Ph-3*)
LA 1589	*Solanum pimpinellifolium*	Wild	Peru	Genome sequenced, widely used for genetic studies
PI 128216	*Solanum pimpinellifolium*	Wild	Bolivia	Resistance source for bacterial spot and bacterial speck
LA 0373	*Solanum pimpinellifolium*	Wild	Peru	Core collection
LA 0400	*Solanum pimpinellifolium*	Wild	Peru	Core collection
LA 0722	*Solanum pimpinellifolium*	Wild	Peru	Core collection
LA 1582	*Solanum pimpinellifolium*	Wild	Peru	Core collection
LA 2181	*Solanum pimpinellifolium*	Wild	Peru	Core collection
Heinz 1706	*Solanum lycopersicum*	Processing	USA	Genome sequenced
OH 88119	*Solanum lycopersicum*	Processing	USA	Early fruit set
OH 9242	*Solanum lycopersicum*	Processing	USA	High lycopene
Liger 87-5	*Solanum lycopersicum*	Processing	China	Current major variety in China
M 82	*Solanum lycopersicum*	Processing	Israel	Widely used in genetic studies
Money maker	*Solanum lycopersicum*	Greenhouse	USA	Widely used in genetic studies
Fla.7600	*Solanum lycopersicum*	Fresh market	USA	Variety with multiple disease resistance genes
Baiguoqiangfeng	*Solanum lycopersicum*	Fresh market	China	Previous major variety in China
Shijifeng	*Solanum lycopersicum*	Fresh market	China	Previous major variety in China
Zhongshu 5	*Solanum lycopersicum*	Fresh market	China	Previous major variety in China
Black cherry	*Solanum lycopersicum* var. *cerasiforme*	Cherry	USA	Brown fruit
LA 1310	*Solanum lycopersicum* var. *cerasiforme*	Cherry	Peru	Salt tolerance
LA 4133	*Solanum lycopersicum* var. *cerasiforme*	Cherry	USA	Core collection, salt tolerance
PI 114490	*Solanum lycopersicum* var. *cerasiforme*	Cherry	UK	Yellow fruit, resistance to bacterial spot

Prediction of InDels between LA 1589 and Heinz 1706

The genomic DNA sequences of S. pimpinellifolium accession LA 1589 (Spimpinellifolium_genome.contigs.fasta.gz) and S. lycopersicum variety Heinz 1706 (S_lycopersicum_chromosomes.2.40.fa.gz) were downloaded to a local computer from the SOL Genomics Network (SGN,http://solgenomics.net/, 19 February 2014, date last accessed). The genomic DNA sequence contigs of LA 1589 were assigned to Heinz 1706 genome using local MUGSY37 downloaded from Sourceforge (http://mugsy.sourceforge.net/, 19 February 2014, date last accessed). InDel polymorphisms referring to Heinz 1706 were mined from the alignments using custom PERL scripts. Flanking sequences of 100 bp from each side of candidate InDels were extracted from Heinz 1706 sequences for insertion and LA 1589 sequences for deletion. The flanking sequences were then blasted against LA 1589 sequences for deletion or Heinz 1706 sequences for insertion using local BLASTall with an E-value of e−20 to remove hits with low similarity. The types (insertion or deletion), lengths, nucleotides, and chromosomal positions of InDels were extracted using a PERL script with the highest score of blast search.

Selection of InDels for validation and primer design

Our initial goal was to verify 3000 candidate InDels of 2 bp or longer evenly distributing on 12 chromosomes. Based on the genome sequenced for Heinz 1706 (760 Mb),34 the average distance between two adjacent InDels would be ~250 kb. The number of InDels to be validated was determined by the length of each chromosome (Table 2). However, we found that the InDels were not always evenly distributed on chromosomes and hotspots have high levels of InDels than other regions. Therefore, we tried to acquire an InDel per 200 kb in each chromosome using a PERL script. If a region on a chromosome did not have InDel variation, the PERL script would make 200 plus 100 kb on circulation until it matched.

Table 2. Summary statistics for primer design, PCR amplification, and polymorphisms

Chromosome	Seqeunce length (~Mb)[a]	No. of primers designed	No. of primers without PCR amplification	No. of primers without polymorphism	No. of primers examined	No. (percentage) of polymorphic InDels		
						S. pimpinellifolium	*S. lycopersicum*var. *cerasiforme*	*S. lycopersicum*
chr01	90.3	362	63	89	210	132 (62.9)	38 (18.1)	22 (10.5)
chr02	49.9	207	10	22	175	98 (56.0)	134 (76.6)	75 (42.9)
chr03	64.8	254	19	31	204	123 (60.3)	128 (62.7)	32 (15.7)
chr04	64.1	254	12	24	218	120 (55.0)	128 (58.7)	144 (66.1)
chr05	65.0	262	33	53	176	112 (63.6)	125 (71.0)	127 (72.2)
chr06	46.0	181	20	40	121	99 (81.8)	83 (68.6)	73 (60.3)
chr07	65.3	259	39	15	205	94 (45.9)	26 (12.7)	17 (8.3)
chr08	63.0	252	23	21	208	160 (76.9)	12 (5.8)	11 (5.3)
chr09	67.7	267	21	34	212	135 (63.7)	80 (37.7)	66 (31.1)
chr10	64.8	255	14	35	206	131 (63.6)	111 (53.9)	11 (5.3)
chr11	53.4	214	8	35	171	80 (46.8)	123 (71.9)	109 (63.7)
chr12	65.5	262	10	86	166	116 (69.9)	40 (24.1)	30 (18.1)
Total	759.8	3029	272	485	2272	1400 (61.6)	1028 (45.2)	717 (31.6)

• aThe sequenced genome size was obtained from Sato et al.34

To design primers for PCR validation of InDels, flanking sequences of 100 bp for each side of candidate InDels were extracted. Primers were designed using local Primer338 downloaded from Sourceforge (http://sourceforge.net/project/showfiles.php?group_id=112461, 19 February 2014, date last accessed) with PCR product length 100–200 bp and the optimal length of primer sequence of 20 bp. Several primer pairs were designed for each InDel. The best primer pair was selected based on the optimal GC content of 40–60% and the difference of GC content between forward and reverse primers <10%. All the process was carried out using custom PERL scripts. Primers were synthesized at Sunbiotech Company (Beijing, China) or Sangong Company (Beijing, China).

Validation of InDels using PCR

The PCR technique was adapted to validate the candidate InDels. All synthesized primers were first used to amplify genomic DNA of tomato lines LA 1589 and Heinz 1706. Only primers that successfully amplified a product and had length polymorphisms were then used to detect polymorphisms in the 22 tomato genotypes.

All PCRs were done in 10-µl reaction volume using the method described in Wei et al.39 Reactions were heated at 95°C for 5 min, followed by 32 cycles of 30 s at 95°C, 30 s at 50–60°C depending on the Tm values of primer pairs, and 30 s at 72°C, with a final extension of 5 min at 72°C. The PCR products were subsequently separated in 8% polyacrylamide gel and visualized using the silver-staining approach.17

Data collection and analysis

The presence or absence of each allele for each InDel was coded by 1 or 0, respectively, and scored for a binary data matrix. Allele frequency of each InDel marker was calculated for each genotype. Nei's genetic distance40was calculated for each pair of tomato genotypes using the programme in the software package PHYLIP 3.695 (http://evolution.genetics.washington.edu/phylip.html, 19 February 2014, date last accessed). An Unweighted Pair Group Method with Arithmetic Mean (UPGMA) cluster analysis was performed to develop a dendrogram.

The occurrences of InDels in coding regions of genes were examined by blasting the flanking sequences of 100 bp for each side of the InDel against the tomato ITAG2.3_cds.fasta downloaded from SGN using a PERL script.

Result

Candidate InDels between LA 1589 and Heinz 1706

A total of 145 695 candidate InDels were identified between the genome sequences of Heinz 1706 and LA 1589, of which 65 619 were insertions and 80 076 were deletions in Heinz 1706 (Table 3). The average size of predicted InDels was 4.1 bp with a range of 1–94 bp, of which ~54.0% were 1 bp, 42.3% were 2–20 bp, and 3.7% were longer than 20 bp. The average density of InDels was one per 5.22 kb with a range of 4.33–6.72 kb on 12 chromosomes. The highest density was on chromosome 6 and the lowest density was on chromosome 12 (Table 3). The least difference of numbers for InDels between 1 bp and >1 bp was observed on chromosome 2 (101), while the largest was on chromosome 10 (1496).

Table 3. Predicted number and frequency of InDels between Heinz 1706 and LA 1589

Chromo-some	No. of predicted InDels			Frequency of InDels (kb/InDel)		
	Total	**1 bp**	**>1 bp**	**Total**	**1 bp**	**>1 bp**
chr01	16 547	8777	7770	5.46	10.29	11.62
chr02	10 695	5398	5297	4.67	9.24	9.42
chr03	12 842	6779	6063	5.05	9.56	10.69
chr04	11 495	6112	5383	5.58	10.49	11.91
chr05	12 148	6816	5332	5.35	9.54	12.19
chr06	10 619	5540	5079	4.33	8.30	9.06
chr07	13 426	7386	6040	4.86	8.84	10.81
chr08	13 776	7591	6185	4.57	8.30	10.19
chr09	11 417	6251	5166	5.93	10.83	13.10
chr10	13 390	7443	5947	4.84	8.71	10.90
chr11	9587	5221	4366	5.57	10.23	12.23
chr12	9753	5432	4321	6.72	12.06	15.16
Total	145 695	78 746	66 949			
Average				5.22	9.65	11.35

Online ISSN 1756-1663 - Print ISSN 1340-2838

Number of primers designed and success of PCR amplification

Using the approach described in the section 'Selection of InDels for validation and primer design' of Materials and methods, 3029 candidate InDels were selected and primers were designed for PCR validation (Supplementary Table S1). The average physical distance between two adjacent InDels was 250 kb with a range of 241 (chromosome 2) to 255 kb (chromosome 3) on 12 chromosomes. PCR results showed that 272 primer pairs could not generate PCR products from the genomic DNA of both Heinz 1706 and LA 1589 (Table 2) . The PCR success rate was 91.0%, which was consistent with our previous finding of 91.9% for PCR amplification of genomic DNA in tomato.9 The InDel sizes of PCR products amplified by most primer pairs (98.5%) were as predicted. However, 23 primer pairs showed smaller and 10 primer pairs showed larger sizes than predicted (Supplementary Table S1). In addition, 485 primer pairs did not show detectable polymorphisms between Heinz 1706 and LA 1589 (Table 2). The InDel sizes between 6 and 30 bp had a high percentage (83.6%) of polymorphism validation, while InDels with sizes of <6 bp and >30 bp received 78.3 or 43.3% polymorphism validation, respectively. Particularly, only one of five InDels was validated when the size was >50 bp (Supplementary Table S2). The primer pairs with PCR failure or non-detectable polymorphisms were excluded, and the remaining 2272 primer pairs were used for subsequent analysis. Therefore, the actual average distance between two adjacent InDels was 334 kb with a range of 285 (chromosome 2) to 430 kb (chromosome 1) on 12 chromosomes.

The 2272 InDel markers generated 5025 alleles in the whole collection of 22 tomato genotypes. The number of alleles generated for all InDels varied from 2 to 8 with an average of 2.2. Among the polymorphic InDels, most (85.3%) had two alleles, 10.7% had three alleles, and 2.7% had four alleles (Fig. 1). Only three and two markers had seven and eight alleles, respectively. Similarly, 84.9% polymorphic InDels in S. pimpinellifolium, 94.7% in S. lycopersicum var. cerasiforme, and 95.8% in S. lycopersicumhad two alleles (Fig. 1).

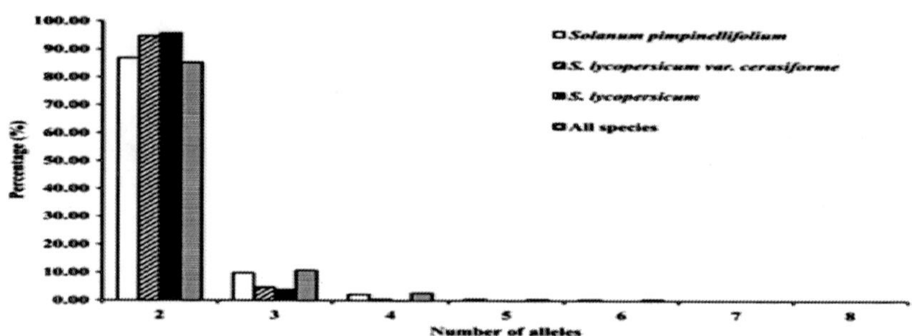

Figure 1 Frequency distribution of InDels (≥2 bp) in Solanum pimpinellifolium, S. lycopersicum var. cerasiforme, and S. lycopersicum.

Marker polymorphisms and distribution among three tomato species

Of the 5025 alleles amplified by 2272 InDel markers, 1930 were shared by all three species. The total number of alleles in each species reduced from 3941 in S. pimpinellifolium to 3431 in S. lycopersicum var. cerasiformeand 3110 in S. lycopersicum (Fig. 2). The number of alleles unique to each species also dramatically decreased from 1382 in S. pimpinellifolium to 56 in S. lycopersicum var. cerasiforme and 60 in S. lycopersicum. Solanum pimpinellifolium shared more alleles with S. lycopersicum var. cerasiformethan with S. lycopersicum.

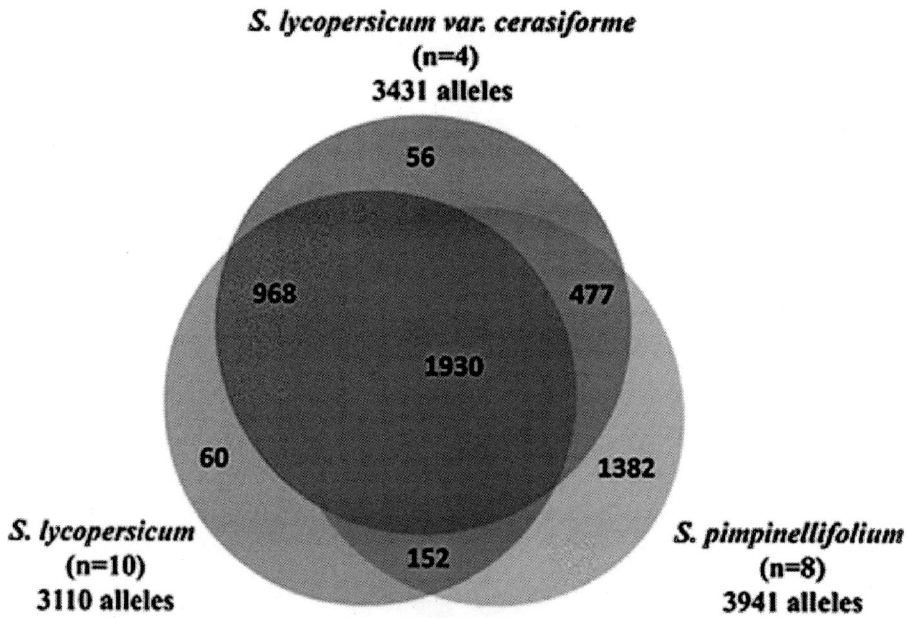

Figure 2 Venn diagram shows the proportion of common alleles amongSolanum pimpinellifolium, S. lycopersicum var. cerasiforme, andS. lycopersicum.

Pairwise comparisons revealed that almost all InDel markers were polymorphic between S. pimpinellifolium and S. lycopersicum var.cerasiforme or S. lycopersicum. However, the proportion of polymorphic InDels reduced to 53.0% between S. lycopersicum var. cerasiforme and S. lycopersicum. There were 0.1–20.7% InDels had alleles alternatively fixed in paired species. In addition, 18.5–26.9% InDels had alleles shared by paired species. Proportions of InDels with alleles specific to one certain species varied from 6.1 to 44.0% (Fig. 3). The proportion of polymorphic InDels was 61.4–100.0% (average 84.6%) between any accession in S. pimpinellifolium and any genotype in S. lycopersicum, 55.3–93.8% (average 71.5%) between any accession in S. pimpinellifolium and any line in S. lycopersicum var. cerasiforme, and 7.7–33.9% (average 19.2%) between any line in S. lycopersicum var. cerasiforme and any genotype inS. lycopersicum
.

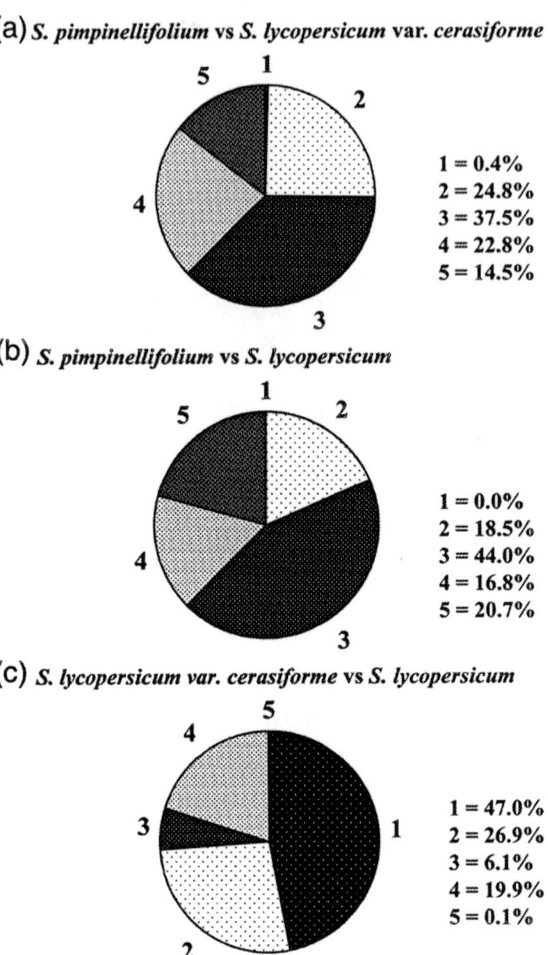

Figure 3. Pairwise comparisons of allelic variation amongSolanum pimpinellifolium, S. lycopersicum var.cerasiforme, and S. lycopersicum. Pie diagrams show the proportion of 2272 InDels that fell into five categories: (1) InDels where a monomorphic allele was shared by all members in the two species; (2) InDels where alleles were found among the members of the two species; (3) InDels where a unique allele was found among members of the first species listed, whereas an alternative allele (found in both groups) was fixed in the second species; (4) InDels where a unique allele was found among members of the second species listed, whereas an alternative allele (found in both species) was fixed in the first species; (5) InDels where the two species were fixed for alternative alleles.

Although the 2272 InDels almost evenly distributed across all 12 chromosomes, the distribution of polymorphic markers varied for three species. Solanum pimpinellifolium had a relatively even distribution of polymorphic InDels on all 12 chromosomes. Solanum lycopersicum var. cerasiforme had the similar distribution pattern of polymorphic InDels as S. pimpinellifoliumon chromosomes 2, 3, 4, 5, 6, 9, 10, and 11, but clusters of polymorphic InDels occurred at some regions on chromosomes 1, 7, and 12. The distribution of polymorphic InDels varied across and within chromosomes in S. lycopersicum. Among six chromosomes with less polymorphic InDels, chromosomes 1, 8, 10, and 12 had relatively even distribution, while the long-arm ends of chromosomes 3 and 7 had more InDels than other regions. There were less InDels at one end of chromosomes 2, 4, 5, 9, and 11. However, chromosomes 5, 9, and 11 showed relatively even distribution. On chromosome 6, the short arm had more polymorphic InDels than the long arm.

The proportion of polymorphic InDels on 12 chromosomes ranged from 45.9 to 81.8% in S. pimpinellifolium, 5.8 to 76.6% in S. lycopersicum var.cerasiforme, and 5.3 to 72.2% in S. lycopersicum (Fig. 4). The numbers of polymorphic InDels considerably decreased on four chromosomes 1, 7, 8, and 12 in S. lycopersicum var. cerasiforme and S. lycopersicum (Table 2). Furthermore, the proportions of polymorphic InDels on chromosomes 3 and 10 were close between S. pimpinellifolium and S. lycopersicum var. cerasiforme, but significantly decreased in S. lycopersicum (Fig. 4). Interestingly, increases of InDel polymorphisms were observed on chromosomes 4, 5, and 11 in S. lycopersicum var. cerasiforme and S. lycopersicum. The proportions of polymorphic InDels also increased on chromosomes 2 and 3 in S. lycopersicum var. cerasiforme.

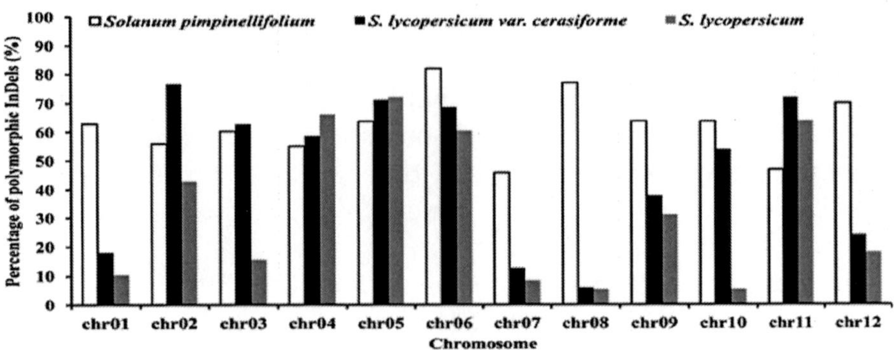

Figure 4. Distribution of the proportion of polymorphic InDels on 12 chromosomes in Solanum pimpinellifolium,S. lycopersicum var.cerasiforme, and S. lycopersicum.

Marker polymorphisms and genetic vitiation within three tomato species

The proportion of polymorphic InDels was 61.6% in 8 S. pimpinellifoliumaccessions, 45.2% in 4 S. lycopersicum var. cerasiforme accessions, and 31.6% in 10 cultivated tomato varieties (Table 2). However, the rate of polymorphic InDels between any two genotypes was low with a range of 14.3–33.6% in S. pimpinellifolium, 17.5–31.5% in S. lycopersicum var.cerasiforme, and 1.5–19.8% in S. lycopersicum (Supplementary Table S3).

Not surprisingly, the eight accessions of S. pimpinellifolium had the largest genetic variation among three species. The average genetic distance was 0.216 with a range from 0.178 (PI 128216) to 0.244 (LA1589). Accessions LA 1589 and LA 2181 had the greatest genetic distance with 0.394, whereas accessions PI 128216 and LA 0373 had the least genetic distance with 0.137. The average genetic distance slightly reduced to 0.202 with a range from 0.162 (LA 4133) to 0.237 (PI 114490) in four S. lycopersicum var. cerasiforme lines, but significantly decreased to 0.108 with a range of 0.086 (Baiguoqiangfeng) to 0.139 (M 82) in 10 varieties of S. lycopersicum. The minimum genetic distance was 0.012 between varieties Liger 87-5 and M 82, followed by 0.015 between varieties Baiguoqiangfeng and Zhongshu 5, while the largest genetic distance was 0.214 between

Shijifeng and M 82.

The dendrogram was constructed from the pairwise genetic distance matrices based on Nei's distance for 22 genotypes. Two distinct groups, A and B, were obtained (Fig. 5). All 8 accessions of S. pimpinellifolium were in Group A, and 10 S. lycopersicum var. cerasiforme cultivars and 4 S. lycopersicum var. cerasiforme accessions were in Group B. The four fresh market cultivars clustered together. However, five processing varieties, one greenhouse variety, and four S. lycopersicum var. cerasiformeaccessions did not form their own clades. Of the four S. lycopersicum var.cerasiforme lines, LA 4133 clustered to three processing and one greenhouse varieties, Black cherry clustered to two processing varieties, while PI 114490 and LA 1310 stood alone.

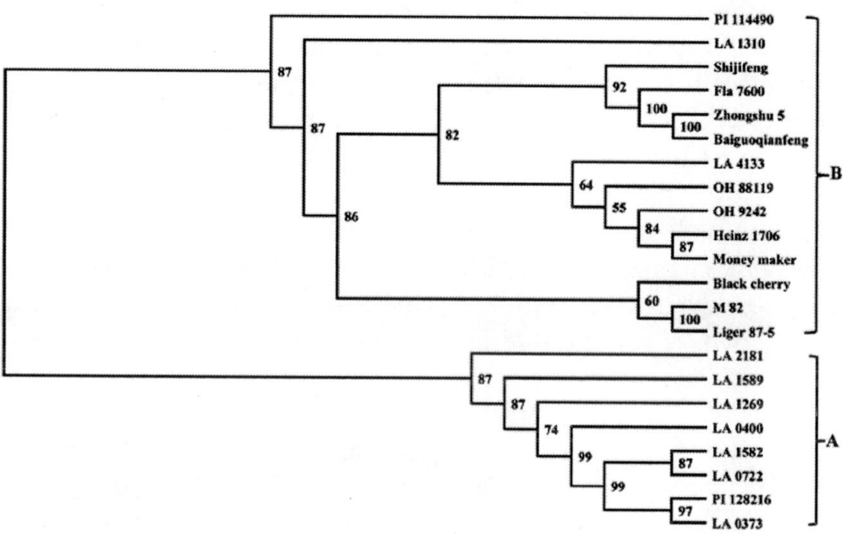

Figure 5. The dendrogram of 22 tomato genotypes based on 2272 InDel marker data, and generated from Nei's genetic distance matrix by UPGMA in PHYLIP 3.695.

Genes with InDels in theCoding Region

Blast search of flanking sequences of 2272 validated InDels against the tomato ITAG2.3_cds.fasta data identified 56 InDels in coding regions of annotated genes), of which 64.3% were deletions in Heinz 1706 and 35.7% were insertion in Heinz 1706. Based on the sizes of InDels, 28.6% of InDels were frame-shift mutations, because the numbers of nucleotides in the InDels were indivisible by three. The remaining 71.4% InDels did not result in frame-shift, but would cause insertion or deletion of some amino acids.

Discussion

Molecular markers are important to genetic study and marker-assisted selection. Large-scale discovery combining high-throughput genotyping of SNPs have shown its power in gene identification and breeding in tomato.12 However, high costs and technical or equipment demands will still be a major obstacle for large-scale use of SNPs in the developing countries.41,42 On the contrary, the genotyping of short InDels is relatively easy and inexpensive with a simple PCR and electrophoresis. Short InDels can also be analysed with high-throughput technologies26,43,44 and in large-scale multiplexing.45 As a type of genetic markers, InDels have been successfully used for forensic analysis46–48 and individual identification44,45 in human, as well as genetic studies in several plant species including rice, wheat, citrus, and Arabidopsis.33 Although the tomato genome sequences have been widely used in various purposes including SNP discovery, genetic mapping, gene prediction, gene expression, genetic diversity, comparative genomics, and epigenetics since their release,49 identification of InDels has so far been confined to detect polymorphisms between wild species and cultivated tomato.34,35 In this study, we identified InDels by comparative analysis of genome sequences between S. pimpinellifolium and S. lycopersicum, and then validated them in 10 cultivated tomato lines via PCR amplification. Of 2272 InDels polymorphic between LA 1589 and Heinz 1706, 31.6% were polymorphic among the 10 cultivated tomato varieties and 1.5–19.8% were polymorphic between any 2 of the 10 cultivated tomato varieties. Based on the total number of InDels (145 695) between LA 1589 and Heinz 1706, we estimated that there were 2100–28 800 InDels between any two cultivated tomato varieties, suggesting that there were abundant InDels for genetic study and marker-assisted selection in the cultivated tomato.

Precise identification of InDels in sequence databases depends on the strategy and the parameters used for data mining as well as the quality of sequence data. Since InDels are the dominant error type generated by 454 pyrosequencing50 and an InDel error rate of one per 6.4 kb was observed in tomato,34 the initial work on identification of InDels between the genomes of LA 1589 and Heinz 1706 did not count InDels of 1 and 2 bp to avoid overestimation of small InDels due to sequencing errors.34 Using a bioinformatic pipeline involving various comparative genomics tools, 9474 InDels of 15–100 bp were identified between LA 1589 and Heinz 1706, and >80% could be verified by PCR (Jiang et al. unpublished data, acquired from ftp://ftp.solgenomics.net/maps_and_markers/LippmanZ/, 19 February 2014, date last accessed). In this study, a total of 145 695 InDels were predicted between LA 1589 and Heinz 1796, which was approximate one-fifth of 749 966 InDels Identified in Sato et al.34 The overall frequency of InDels (one per 5.22 kb) was also much lower than one per 110 bp in Sato et al.34 However, the number (9137) of InDels of 15–94 bp was close to the results of Jiang et al., though the strategies used for InDels identification were different. Owing to the lack of methodology description in Sato et al.,34 we were not able to determine the cause of the difference between two studies. Two points might be worthy of notice. First, the lengths of putative InDels identified in two studies were different with ranges of 3–300 bp in Sato et al.34 and 1–94 bp in this study. We could not identify any InDels >94 bp using our methodology. Secondly, the rate of validation (82.4%) was close to 81.7% obtained in Koenig et al.,35 though the comparisons involved in different wild species and cultivated varieties, indicating that ~20% of predicted InDels (≥2 bp) were false due to sequencing error. All these suggested that our prediction might be more close to the real number of InDels in the currently available genome sequences of LA 1589 and Heinz 1706.

The polymorphic InDels evenly distributed across all 12 chromosomes inS. pimpinellifolium, but appeared non-randomly distributed across and within chromosomes in S. lycopersicum var. cerasiforme and S. lycopersicum.

Domestication and selection could be one causal of this difference. For example, there were 38 and 35 polymorphic InDels at the bottom (~11 Mb) of chromosome 2 in S. pimpinellifolium and S. lycopersicum var. cerasiforme, respectively, but only two InDels were polymorphic in S. lycopersicum. This might be due to the existence of quantitative trait loci for fruit weight and selection for large fruit in S. lycopersicum.12 In addition, several studies have proved that the introgression of disease resistance genes in many cultivars has strong influence on SNP patterns.19,51 This kind of introgression could also cause the difference of polymorphic InDels distribution among three species.

It has been suggested that domestication and inbreeding dramatically reduced the genetic variation52 and modern cultivars have less genetic variation than old ones in tomato.53,54 In this study, genetic variation of three species was investigated using the same large set of InDel markers, which allowed us to compare genetic polymorphisms among and within species at the same time. The number of polymorphic InDels, the total number of alleles amplified by InDel markers, and the average genetic distance in 10 S. lycopersicum varieties significantly reduced comparing with those in 8 S. pimpinellifolium accessions, supported the reduction of genetic variation in cultivated tomato. The four S. lycopersicum var.cerasiforme accessions showed an intermediate amount of genetic diversity between S. lycopersicum and S. pimpinellifolium, which was consistent with previous findings.55,56 However, some novel alleles occurred in both S. lycopersicum var. cerasiforme and S. lycopersicum, suggesting that domestication and selection could also generate new variation.

The occurrence of InDels in coding regions of a gene can either cause frame-shift or amino acid InDels, which most likely alternates the gene function and results in phenotype change.57 A Rider mutational insertion event occurring in the first exon of the Psy1 gene causes the early termination of Psy1 transcription that results in yellow flesh in the tomator mutant.58 A single-base deletion mutation in the coding region ofSlIAA9 gene, an Aux/IAA gene involving in tomato leaf morphology, converts tomato compound leaves to simple leaves.59 InDels occurring in the promoter region can also affect the gene expression.60 Here, we identified 145 695 InDels between LA 1589 and Heinz 1706, and 31.6% of them were polymorphic in cultivated tomatoes. The percentage of InDels (2.5%) occurring in coding regions of genes identified in this study was much lower than our recent work (19.7%) on comparative analysis of resistance-like genes between LA 1589 and Heinz 1706.61 Identification of specific genes in our previous work other than a random sample in this study could cause the different proportions of InDels in coding regions.

In conclusion, there are abundant short InDels in cultivated tomato. Identification and validation of this kind of short InDels will not only provide molecular markers for genetic study and marker-assisted selection in breeding, but also provide useful information for gene cloning and functional analysis.

ACKNOWLEDGEMENTS

We thank Dr David M. Francis at the Ohio State University (USA), and Tomato Genetics Resource Center at the University of California (Davis, USA) for providing seeds of some tomato lines used in this study.

REFERENCES

1. Foolad M.R., Panthee R.D. Marker-assisted selection in tomato breeding. Crit. Rev. Plant Sci. 2012;31:93-123.
2. Shirasawa K., Hirakawa H. DNA marker applications to molecular genetics and genomics in tomato. Breeding Sci. 2013;63:21-30.
3. Miller J.C., Tanksley S.D. RFLP analysis of phylogenetic relationships and genetic variation in the genus Lycopersicon. Theor. Appl. Genet. 1990;80:437-48.
4. Yang W.C., Bai X.D., Kabelka E., et al . Discovery of single nucleotide polymorphisms in Lycopersicon esculentum by computer aided analysis of expressed sequence tags. Mol. Breeding 2004;14:21-34.
5. Labate J.A., Baldo A.M. Tomato SNP discovery by EST mining and resequencing.Mol. Breeding 2005;16:343-9.
6. Yamamoto N., Tsugane T., Watanabe M., et al . Expressed sequence tags from the laboratory-grown miniature tomato (Lycopersicon

esculentum) cultivar Micro-Tom and mining for single nucleotide polymorphisms and insertions/deletions in tomato cultivars. Gene 2005;56:127-34.

7. Shirasawa K., Isobe S., Hirakawa H., et al . SNP discovery and linkage map construction in cultivated tomato. DNA Res. 2010;17:381-91.
8. van Deynze A., Stoffel K., Buell C.R., et al . Diversity in conserved genes in tomato. BMC Genomics 2007;8:465.
9. Wang Y.Y., Chen J., Francis D.M., Shen H.L., Wu T.T., Yang W.C. Discovery of intron polymorphisms in cultivated tomato using both tomato and Arabidopsis genomic information. Theor. Appl. Genet. 2010;121:1199-207.
10. Sim S.-C., Robbins M.D., Chilcott C., Zhu T., Francis D.M. Oligonucleotide array discovery of polymorphisms in cultivated tomato (Solanum lycopersicum L.) reveals patterns of SNP variation associated with breeding. BMC Genomics 2009;10:466.
11. Hamilton J.P., Sim S.C., Stoffel K., Van Deynze A., Buell C.R., Francis D.M. Single nucleotide polymorphism discovery in cultivated tomato via sequencing by synthesis.Plant Genome 2012;5:17-29.
12. Víquez-Zamora M., Vosman B., van de Geest H., et al . Tomato breeding in the genomics era: insights from a SNP array. BMC Genomics 2013;14:354.
13. Shirasawa K., Fukuoka H., Matsunaga H., et al . Genome-wide association studies using single nucleotide polymorphism markers developed by re-sequencing of the genomes of cultivated tomato. DNA Res. 2013;20:593-603.
14. Robbins M.D., Sim S.C., Yang W.C., et al . Mapping and linkage disequilibrium analysis with a genome-wide collection of SNPs that detect polymorphism in cultivated tomato. J. Exp. Bot. 2011;62:1831-45.
15. Hirakawa H., Shirasawa K., Ohyama A., et al . Genome-wide SNP genotyping to infer the effects on gene functions in tomato. DNA Res. 2013;20:221-33.
16. Sim S.-C., Durstewitz G., Plieske J., et al . Development of a large SNP genotyping array and generation of high-density genetic maps in tomato. PLoS ONE2012;7:e40563.
17. Chen J., Wang H., Shen H.L., et al . Genetic variation in tomato populations from four breeding programs revealed by single nucleotide polymorphism and simple sequence repeat markers. Sci. Hortic. 2009;122:6-16.
18. Sim S.-C., Robbins M.D., van Deynze A., Michel A.P., Francis D.M.

Population structure and genetic differentiation associated with breeding history and selection in tomato (Solanum lycopersicum L.). Heredity 2011;106:927-35.

19. Sim S.-C., van Deynze A., Stoffel K., et al . High-density SNP genotyping of tomato (Solanum lycopersicum L.) reveals patterns of genetic variation due to breeding. PLoS ONE 2012;7:e45520.
20. Corrado G., Piffanelli P., Caramante M., Coppola M., Rao R. SNP genotyping reveals genetic diversity between cultivated landraces and contemporary varieties of tomato. BMC Genomics 2013;14:835.
21. Collins F.S., Drumm M.L., Cole J.L., Lockwood W.K., Vande Woude G.F., IannuzziM.C. Construction of a general human chromosome jumping library, with application to cystic fibrosis. Science 1987;235:1046-9.
22. Usdin K. The biological effects of simple tandem repeats: lessons from the repeat expansion diseases. Genome Res. 2008;18:1011-9.
23. MacArthur D.G., Tyler-Smith C. Loss-of-function variants in the genomes of healthy humans. Hum. Mol. Genet. 2010;19:R125-130.
24. Stenson P.D., Ball E.V., Howells K., Phillips A.D., Mort M., Cooper D.N. The Human Gene Mutation Database: providing a comprehensive central mutation database for molecular diagnostics and personalized genomics. Hum. Genomics 2009;4:69-72.
25. Montgomery S.B., Goode D.L., Kvikstad E., et al . The origin, evolution, and functional impact of short insertion-deletion variants identified in 179 human genomes. Genome Res. 2013;23:749-61.
26. Mills R.E., Luttig C.T., Larkins C.E., et al . An initial map of insertion and deletion (INDEL) variation in the human genome. Genome Res. 2006;16:1182-90.
27. Mullaney J.M., Mills R.E., Pittard W.S., Devine S.E. Small insertions and deletions (INDELs) in human genomes. Hum. Mol. Genet. 2010;19:R131-136.
28. Bromberg Y. Building a genome analysis pipeline to predict disease risk and prevent disease. J. Mol. Biol. 2013;425:3993-4005.
29. Hou X., Li L., Peng Z., et al . A platform of high-density INDEL/CAPS markers for map-based cloning in Arabidopsis. Plant J. 2010;63:880-8.
30. Pacurar D.I., Pacurar M.L., Street N., et al . A collection of INDEL markers for map-based cloning in seven Arabidopsis accessions. J. Exp. Bot. 2012;63:2491-501.
31. Shen Y.J., Jiang H., Jin J.P., et al . Development of genome-wide DNA polymorphism database for map-based cloning of rice genes. Plant Physiol. 2004;135:1198-205.

32. Zeng Y.X., Wen Z.H., Ma L.Y., Ji Z.J., Li X.M., Yang C.D. Development of 1047 insertion-deletion markers for rice genetic studies and breeding. Genet. Mol. Res.2012;12:5226-35.

33. Wu D.H., Wu H.P., Wang C.S., Tseng H.Y., Hwu K.K. Genome-wide InDel marker system for application in rice breeding and mapping studies. Euphytica2013;192:131-43.

34. Sato S., Tabata S., Hirakawa H., et al . The tomato genome sequence provides insights into fleshy fruit evolution. Nature 2012;485:635-41.

35. Koenig D., Jimenez-Gomez J.M., Kimura S., et al . Comparative transcriptomics reveals patterns of selection in domesticated and wild tomato. Proc. Natl. Acad. Sci. USA 2013;110:E2655-62.

36. Kabelka E., Franchino B., Francis D.M. Two loci from Lycopersicon hirsutumLA407 confer resistance to strains of Clavibacter michiganensis subsp.Michiganensis. Phytopathology 2002;92:504-10.

37. Angiuoli S.V., Salzberg S.L. Mugsy: fast multiple alignment of closely related whole genomes. Bioinformatics 2011;27:334-42.

38. Rozen S., Skaletsky H. Primer3 on the WWW for general users and for biologist programmers. Methods Mol. Biol. 2000;132:365-86.

39. Wei J.L., Hu X.R., Yang J.J., Yang W.C. Identification of single-copy orthologous genes between Physalis and Solanum lycopersicum and analysis of genetic diversity in Physalis using molecular markers. PLoS ONE 2012;7:e50164.

40. Nei M. Genetic distance between populations. Am. Nat. 1972;106:283-92.

41. Ribaut J.M., de Vicente M.C., Delannay X. Molecular breeding in developing countries: challenges and perspectives. Curr. Opin. Plant Biol. 2010;13:1-6.

42. Jiang G.L. Plant marker-assisted breeding and conventional breeding: challenges and perspectives. Adv. Crop Sci. Technol. 2013;1:e106.

43. Weber J.L., David D., Heil J., Fan Y., Zhao C., Marth G. Human diallelic insertion/deletion polymorphisms. Am. J. Hum. Genet. 2002;71:854-62.

44. Oka K., Asari M., Omura T., et al . Genotyping of 38 insertion/deletion polymorphisms for human identification using universal fluorescent PCR. Mol. Cell. Probes 2014;28:13-8.

45. Pereira R., Phillips C., Alves C., Amorim A., Carracedo A., Gusmão L. A new multiplex for human identification using insertion/deletion polymorphisms.Electrophoresis 2009;30:3682-90.

46. Li C.T., Zhang S.H., Zhao S.M. Genetic analysis of 30 InDel markers for forensic use in five different Chinese populations. Genet. Mol. Res.

2011;10:964-79.

47. Li C., Zhao S., Zhang S., et al. Genetic polymorphism of 29 highly informative InDel markers for forensic use in the Chinese Han population. Forensic Sci. Int. Genet.2011;5:e27-30.

48. Fondevila M., Phillips C., Santos C., et al . Forensic performance of two insertion-deletion marker assays. Int. J. Legal Med. 2012;126:725-37.

49. Menda N., Strickler S.R., Mueller L.A . Advances in tomato research in the post-genome era. Plant Biotechnol. 2013;30:243-56.

50. Shendure J., Ji H. Next-generation DNA sequencing. Nat. Biotechnol.2008;26:1135-45.

51. Causse M., Desplat M., Pascual L., et al . Whole genome resequencing in tomato reveals variation associated with introgression and breeding events. BMC Genomics2013;14:791.

52. Bai Y.L., Lindhout P. Domestication and breeding of tomatoes: what have we gained and what can we gain in the future? Ann. Bot. 2007;100:1085-94.

53. Archak S., Karihaloo J.L., Jain A. RAPD markers reveal narrowing genetic base of Indian tomato cultivars. Curr. Sci. 2002;82:1139-43.

54. Hu X.R., Wang H., Chen J., Yang W.C. Genetic diversity of Argentina tomato varieties revealed by morphological traits, simple sequence repeat, and single nucleotide polymorphism markers. Pak. J. Bot. 2012;44:485-92.

55. Rick C.M., Fobes F. Allozyme variation in the cultivated tomato and closely related species. Bull. Torr. Bot. Club 1975;102:376-84.

56. Ranc N., Muños S., Santoni S., Causse M. A clarified position for Solanum lycopersicum var. cerasiforme in the evolutionary history of tomatoes (solanaceae).BMC Plant Biol. 2008;8:130.

57. Hu J., Ng P.C. SIFT Indel: predictions for the functional effects of amino acid insertions/deletions in proteins. PLoS ONE 2013;8:e77940.

58. Cheng X.D., Zhang D.F., Cheng Z.K., Keller B., Ling H.Q. A new family of Ty1-copia-like retrotransposons originated in the tomato genome by a recent horizontal transfer event. Genetics 2009;181:1183-93.

59. Zhang J., Chen R., Xiao J., et al . A single-base deletion mutation in SlIAA9 gene causes tomato (Solanum lycopersicum) entire mutant. J. Plant Res. 2007;120:671-8.

60. Ruíz-Rivero O.J., Prat S. A-308 deletion of the tomato LAP promoters is able to direct flower-specific and MeJA-induced expression in transgenic plants. Plant Mol. Biol. 1998;36:639-48.

61. Ni X., Yang J., Sun S., Yang W. Identification and analysis of resistance-like genes in the tomato genome. J. Phytopathol. 2013;162:137-46

Chapter 9

FROM CROP DOMESTICATION TO SUPER-DOMESTICATION

D. A. Vaughan[1],*, E. Balázs[2] and J. S. Heslop-Harrison[3]

[1]National Institute of Agrobiological Sciences, Kannondai 2-1-2, Tsukuba 305-8602, Ibaraki, Japan

[2]Agriculture Research Institute, Department of Applied Genomics, Hungarian Academy of Sciences, Brunszvik u2, Martonvasar 2462, Hungary

[3]Department of Biology, University of Leicester, Leicester LE1 7RH, UK

ABSTRACT

Research related to crop domestication has been transformed by technologies and discoveries in the genome sciences as well as information-related sciences that are providing new tools for bioinformatics and systems' biology. Rapid progress in archaeobotany and ethnobotany are also contributing new knowledge to understanding crop domestication. This sense of rapid progress is encapsulated in this Special Issue, which contains 18 papers by scientists in botanical, crop sciences and related disciplines on the topic of crop domestication. One paper focuses on current themes in the genetics of crop domestication across crops, whereas other papers have a crop or geographic focus. One feature of progress in the sciences related to crop domestication is the availability of well-characterized germplasm resources in the global network of genetic resources centres (genebanks). Germplasm in genebanks is providing research materials for understanding domestication as well as for plant breeding. In this review, we highlight current genetic themes related to crop domestication. Impressive progress in this field in recent years is transforming plant breeding into crop engineering to meet the human need for increased crop yield with the minimum environmental impact – we consider this to be 'super-domestication'. While the time scale of domestication of 10 000 years or less is a very short evolutionary time span, the details emerging of what has happened and what is happeningprovide a window to see where domestication might – and can – advance in the future.

INTRODUCTION

As human societies have evolved so have the plants in the human environment. The transition from gathering wild plants to cultivation involved increasing interaction between humans and the plants they used. The subsequent genetic changes in these plants resulting in domestication of some of these cultivated species reflect the genius of early farmers, who were the first plant breeders. The present generation of plant breeders has tools available that enable them to be plant engineers. In this article that overviews the topic and the range of papers in this Special Issue of Annals of Botany on crop domestication, we discuss current themes related to the genetics of crop domestication and crop genepools that are helping to accelerate the transition from plant breeding to plant engineering, from crop domestication to crop super-domestication.

Cultivated or domesticated plants, when mankind propagates, plants and harvests them, have played significant roles in many of the advances that pure and applied botanical sciences have made in the last few centuries. The earliest farmers recognized useful genetic variation that could be chosen from the wild, planted, harvested and reselected in order to gradually develop improved populations with a range of desirable traits. The domesticated forms bore only limited resemblance to their wild ancestors due to the selection of domestication genes. Early in the 18th century, the first conscious hybridization of plants occurred using the cultivated ornamental species Dianthus (Phillips, 2006). Knowledge of plant hybridization paved the way for the use by plant breeders of intra- and interspecific hybridization in crop improvement. Mendel›s discovery of the laws of genetics, based most famously on his experiments with the domesticated garden pea, led to improved understanding of the variation in domesticated and wild species that so impressed Darwin (1859). Subsequently, sophisticated crop breeding programmes were developed that enabled more efficient introduction of desirable traits from one cultivar to another. Knowledge of the evolutionary relationships between crops and their wild progenitors has facilitated more efficient exploitation of the genetic resources represented by the wild relatives of domesticated species (for a review, see Hajjar and Hodgkin, 2007). Currently, domesticated crops such as rice, maize and tomato are major targets of studies in molecular genetics. Research on domestication-related traits is leading to a better understanding of how genetic control of phenotypic differences is effected. For example, in determining the extent to which similar (orthologous) genes are involved in producing similar phenotypes in distantly related species and working out how to transfer desirable genes between species that cannot be hybridized sexually, and then how to control the expression of the genes once they are transferred.

Hybridization and selection have both been involved in the origin of crops and the process of domestication since early times. Bread wheat, which is a hexaploid ($2n = 6x = 42$ chromosomes), arose through fortuitous hybridization between tetraploid wheat (Triticum turgidum ssp. dicoccum, $2n = 4x = 28$) and diploid goatgrass (Aegilops tauschii, $2n = 2x = 14$), a weed of early wheat fields. New groups of bananas and plantains developed when diploid domesticated bananas (genome AA) spread into the range of wild Musa balbisiana (genome BB), producing the AAB and ABB triploids (see Heslop-Harrison and Schwarzacher, 2007). Modern strawberries (Fragaria × ananassa) are a consequence of hybridization between North American F. virginiana and South American F. chiloensiswhen these hitherto geographically isolated species were cultivated in close proximity in European gardens

(Bringhurst and Voth, 1984). Selection, both intentionally by humans (conscious selection) and as a result of environmental factors (natural or automatic selection) has, in most crops, established the traits associated with the domestication syndrome (Hammer, 1984). Weed species have co-evolved with crops and have been under similarly intense selection pressures. Weed control through agronomy and the better competitive ability of crops has been continuous over the 10 000 years of agriculture, but weeds still reduce yields and contaminate crops. In rice, changing agricultural practices is leading to the emergence of new weedy rice forms in the last decade (Caoet al., 2006). Thus, as with the crops, weeds are also evolving rapidly under selection.

Analysis of the genetic, genomic and molecular basis of the traits selected by early farmers that constitute the domestication syndrome in crops, such as loss of seed shattering and increased organ size, has been a major focus of much recent research. Profound insights into traits associated with crop domestication have resulted from the technologies and discoveries that make DNA analysis and manipulation possible (for review see Phillips, 2006). These technologies are associated with PCR, development of transgenic crops and chromosome painting, as well as DNA sequencing and information processing. The rapid spread of communication technologies enable new knowledge to be disseminated at a remarkable speed worldwide and have ushered in global research initiatives related to crop genome analysis and related research. These tools have been used in various crops for the development of genome maps, QTL analysis, whole-genome sequencing, fine-resolution mapping and gene cloning. These scientific advances have also contributed to crop improvement, with the application of, for example, marker-assisted selection. The use of molecular techniques has provided a range of new insights into domestication and its future course.

FINE-RESOLUTION MAPPING AND GENE CLONING OF DOMESTICATION-RELATED GENES

Plants have relatively few morphological traits that are controlled by single genes with distinct alleles showing Mendelian segregation

- Mendel was fortunate to find seven independent and non-interacting genes. The development of molecular markers to analyse directly DNA or gene products has meant that tens of thousands of segregating polymorphisms can be analysed in a species. This has made it possible to locate genes controlling both qualitative and quantitative characters very precisely within the genome, understand their interactions, and then to study the phenotypic effects of individual genes or small regions of chromosome in segregating populations (Paterson, 2002; Hancock, 2005; Zhang et al., 2005). In a number of studies, quantitative trait loci (QTL) affecting different elements of the domestication syndrome are collocated to the same locus or the same small chromosomal region, suggesting either pleiotropy or close linkage of the gene(s) concerned. It can be difficult to distinguish pleiotropy from close linkage (Bomblies and Doebley, 2006) but in this Special Issue, Weeden (2007) gives some examples of pleiotropy in pea; other examples are discussed below.

Fine mapping of genes has led to the ability to clone domestication-related genes and unravel the molecular basis of domestication-related changes. For example, the two genes that are most important in relation to spikelet shattering in rice (sh4 and qSH1) have been cloned (Konishi et al., 2006; Li et al., 2006; discussed by Sweeney and McCouch, 2007, in this Special Issue). sh4 is the key shattering gene that distinguishes cultivated from wild rice, while the qSH1 gene controls the difference in the degree of shattering between some indica and japonica varieties of rice. sh4 is a transcription regulator and a single amino acid substitution results in reduced shattering. For qSH1 a single nucleotide in the regulatory region of this gene results in the altered level of seed shattering. sh4 activates the abscission process while qSH1 regulates abscission-layer formation. Sequence analysis of sh4 has revealed a single base-pair mutation that is responsible for non-shattering and this change is the same in both indica and japonica rice varieties (Lin et al., 2007). This result raises doubts about whether Asian rice was domesticated more than once, as has been suggested in several recent papers (for a review see Sang and Ge, 2007). In contrast, sequencing and comparing seven loci in wild and landrace barley have provided strong evidence that barley was domesticated once in the Fertile Crescent and a second time between 1500 and 3000 km to the east (Morrell and Clegg, 2007). Analysis of domestication genes across diverse germplasm can resolve questions about where, from what and how many times a crop was domesticated.

Differences in nucleotide sequence and/or levels of transcription of different alleles of transcriptional regulators affect the phenotypes produced by target genes. In wheat, the Q gene is pleiotropic for many domestication traits. The wild-type allele q is associated with a fragile rachis and grain that does not thresh free of the chaff, whereas the domestication allele Q is associated with a tough rachis and free-threshing grain (Simons et al., 2006). Comparison of the structure and activity of these two alleles suggests that q is transcribed at lower levels than Q and that the q protein functions less efficiently than the protein product of Q (Simons et al., 2006). Q is not known in the wild progenitors of wheat, but human selection post-domestication seems to have resulted in up-regulation of Q such that Q has more than twice the effect of q(Simons et al., 2006). Similarly, some of the differences in branching and spikelet suppression distinguishing domesticated maize from wild teosinte and controlled by tb1 have been attributed to up-regulation oftb1 in maize (Hubbard et al., 2002). To date, most domestication genes that have been cloned are diverse transcription factors that are usually functional (Doebley et al., 2006; Komatsuda et al., 2007). Thus the role of human selection on wild populations during crop domestication at the gene level has been modification rather than elimination of gene function (Consonni et al., 2005; Doebley et al., 2006). This perhaps reflects the relative rarity of mutations leading to new structural or functional genes and the short time span of crop domestication.

DOMESTICATION TRAIT ALLELES CAN BE FOUND IN WILD POPULATIONS

Traits of the domestication syndrome such as loss of seed dispersal, loss of seed dormancy or loss of protection against herbivores are considered disadvantageous in wild plants (e.g. Crawley and Brown, 1995). These are often recessive alleles so their effects would be masked in the heterozygotes that make up the bulk of many wild populations. However, these recessive alleles become exposed through the inbreeding associated with domestication in many crops.

In maize, pleiotropic effects associated with zfl2 are such that selection for increased yield via increases in row number controlled by zfl2 would probably select also for earlier flowering and fewer ears placed lower on the plant (Bomblies and Doebley, 2006). This led Bomblies and Doebley (2006) to suggest that, in general, undesirable secondary effects associated with pleiotropic genes could limit selection for favourable 'domestication alleles' during early stages of the differentiation of a crop from its wild progenitor. On the other hand, selection for beneficial traits controlled by pleiotropic genes could result in associated neutral or even detrimental traits being concurrently selected. This may explain, at least partially, the presence, in wild populations, of alleles for traits of the domestication syndrome that apparently evolved prior to domestication and survived despite their possibly deleterious effects in the wild. Examples of this include alleles of the 'hidden QTL' fw2·2 for increased fruit size in cultivated tomatoes (Solanum lycopersicum) that are also found in the wild cherry tomato (S. lycopersicum var. cerasiforme; Nesbitt and Tanksley, 2002; Bai and Lindhout, 2007). Alleles of the regulatory locus CAULIFLOWER (BoCAL) in Brassica oleracea that contribute to, but are insufficient to cause, development of abnormal inflorescence are present in moderate frequency in wild populations of B. oleracea subsp.oleracea (Purugganan et al., 2000).

A key gene responsible for some differences between maize and its wild progenitor is the teosinte branched 1 (tb1) mutant that has pleiotropic effects on apical dominance, length of lateral branches, growth of blades of leaves on lateral branches, and development of the pedicillate spikelet in the female inflorescence (Hubbard et al., 2002). In the progenitor of maize, teosinte (Zea mays var. parviglumis), a tb1 region haplotype with sequences identical to that of the major maize tb1 haplotype was found. This result suggested that haplotypes that confer maize-like phenotypes could predate domestication (Clark et al., 2004). Thus, the high-speed evolution represented by crop domestication can be the result of strong selection pressures on pre-existing variation.

Humans caused a major shift in the morphological traits of wild plants by selecting genes of both large effect and small effect to create crops with higher yield of desired product. In azuki bean (Vigna angularis) domestication has reduced seed yield on a per plant basis because farmers have selected determinate plants with larger pods and fewer large seeds per pod than its progenitor wild relative

(A. Kaga, NIAS, Japan, unpubl. res.). Mathematical analysis of the functional and structural components of yield, including harvest index – a systems' biology approach – have great potential to indicate future directions for selection (Guo et al., 2006). The wild relatives of crops continue to be an important reservoir of genes for potential use in agriculture. Sometimes, the genes they have furnished have had a dramatic effect on yield, as shown byTanksley and McCouch (1997) and by Cheng et al. (2007) in this Special Issue. Therefore, there is a continued urgency to conserve these wild genetic resources appropriately, both in situ and ex situ, and to characterize them for future crop improvement.

ORTHOLOGUES OF DOMESTICATION GENES AND THEIR ACTION

There are many different families of transcriptional regulators in plants and the transcriptional regulators involved in domestication discussed byDoebley et al. (2006) all belong to different families. Within a given family of transcriptional regulators, gene structure may be sufficiently conserved for similarities to be identified not just between genera of the same plant family but between taxonomically very distantly related species. Thus,monoculm1 in maize shares similarities with LATERAL SUPPRESSOR fromArabidopsis thaliana and tomato (see Doust, 2007, in this Special Issue) and Q in wheat is similar to APETALA2 (AP2) of Arabidopsis (Simons et al., 2006). AP2-like genes appear to have a wide range of roles in plant development, but Q is so far the only AP2-like gene implicated in domestication (Simons et al., 2006). One of the genes affecting shattering in rice, qSH1, may be an orthologue of REPLUMLESS (RPL) in Arabidopsis(Konishi et al., 2006). REPLUMLESS is involved in formation of an abscission layer in the wall of the fruit, whereas qSH1 affects formation of an abscission layer between pedicel and spikelet, but Konishi et al. (2006)suggest that this difference could be explained by differences in the transcriptional control of RPL and qSH1. The duplicate genes zfl1 and zfl2of maize are orthologous to the FLORICAULA/LEAFY (FLO/LFY) genes of species of Antirrhinum and Arabidopsis, amongst others (Bomblies and Doebley, 2006). Among the various effects suggested for these genes is the change in phyllotaxy that produces whorled organs during flower development. In maize, zfl2 is the candidate gene for a major effect QTL controlling the whorled versus two-ranked arrangement of female spikelets in maize versus teosinte. FLO/LFY-like genes have not been reported to affect inflorescence phyllotaxy in any other species, butBomblies and Doebley (2006) suggest that a change in expression pattern could have allowed one of their orthologues to be annexed for a new role in maize.

Rice contains an orthologue of maize tb1, OsTB1, that, like maize tb1, affects lateral branching (Takeda et al., 2003). Transgenic rice carrying an extra dose of OsTB1 produced many fewer tillers than normal because of over-expression of OsTB1. A known mutant, fine culm1 (fc1), with enhanced tiller production, mapped to the same locus as OsTB1, suggesting that fc1 is an allele of OsTB1. Sequencing of fc1 showed a deletion generating a premature stop codon, such that the predicted polypeptide product lacked the domain implicated in the DNA binding activity of the class of transcriptional regulators to which tb1 belongs.Takeda et al. (2003) therefore suggest that alterations in the expression of OsTB1 through dosage effects or use of mutants could be used to increase or decrease tiller number at will and thereby adapt rice morphology to differing agronomic situations (see also Doust, 2007, in this Special Issue).

In the major oilseed crop canola or oilseed rape (Brassica juncea, B. napusand B. rapa) losses of between 10–50 % of yield can occur due to unsynchronized pod shattering (Østergaard et al., 2006) and require extensive management, including spraying with crop dessicants before harvest and windrowing before threshing. Arabidopsis has proved to be a useful model to study the phenomenon, where a transcriptional regulator,FRUITFUL (FUL), mediates pod dehiscence by inhibiting expression of genes controlling shattering. When this transcriptional regulator was introduced in B. juncea it was over-expressed and pods had no shattering. Further fine-tuning of the expression of this gene in canola may enable the required level of post-harvest shattering to be achieved (Østergaard et al., 2006). Such intentional manipulations to fine-tune gene activity will certainly constitute super-domestication, where genetics interacts with crop management and agronomy.

Evidently much remains to be learned about the actions of transcriptional regulators and how they in turn are regulated.

Recently, Clark et al. (2006)located a factor or factors controlling the levels of the message produced by the transcriptional regulator teosinte branched 1 (tb1) in maize, and hence the phenotypic differences between maize and teosinte associated with tb1, to an intergenic region upstream from tb1. This region consists of a mixture of repetitive and unique sequences not previously considered to contribute to phenotypic variation. Doebley and Lukens (1998) had earlier proposed that modifications in cis-regulatory regions of transcriptional regulators would prove a predominant means for the evolution of novel forms, and the findings of Clark et al. (2006) appear to provide a supporting example. Plant-breeding-related companies are already looking at the effects of up- and downregulating all transcription factors in a given genome, aiming to learn more about the target genes of different transcription factors and producing a super-domesticate (Doebley et al., 2006), perhaps with more success than gene mutation as a source of Dobzhansky's 'hopeful monsters'.

EFFECTS OF SELECTION ON DOMESTICATION GENES

Molecular techniques are not just enabling the position of domestication-related genes to be resolved but they can provide information on the effects of selection and number of generations required for domestication. By studying nucleotide polymorphism in different accessions of a crop upstream and downstream from domestication-related genes, it is possible to determine the extent to which selection is acting across the genome, the selective sweep (Clark et al., 2004). Positive directional selection leads to reduced variation and linkage disequilibria in the respective regions (Palaisa et al., 2004). By comparing sequence diversity around a domestication gene in the crop and its progenitor, a new view of the processes that sculptured the formation of the crops species can be attained. By analysis of nucleotide polymorphism around the teosinte branched 1 (tb1) gene in a wide variety of maize accessions, it was found that human selection acted on the gene's regulatory region and was not detected in the protein-coding region (Wang et al., 1999; Clark et al., 2004). This was considered to be a consequence of the high rates of recombination in maize. From the analysis it was estimated that the time taken to domesticate maize was between 315–1023 years

(Wang et al., 1999). Studies of wheat remains at archaeological sites in southern Turkey and Syria, where wheat domestication is believed to have occurred, reveal a gradual change from dehiscent to indehiscent spikelets, suggesting indehiscence took over one millennium to become established (Tanno and Wilcox, 2006). Archaeological remains of rice from the lower Yangtze river suggests that rice domestication was a slow process (seeFuller, 2007, in this Special Issue) and this is supported by the wild-rice harvesting methods used today, which do not provide a selection pressure for non-shattering spikelets (Fig. 1). Both molecular and archaeobotanical studies suggest a long period of gathering and cultivation preceded domestication for these cereals. While domestication represents rapid change in evolutionary terms, in cereals the transition in the suite of characters that changed wild populations into domesticated crops took place over many centuries or millennia.

FIG. 1. Alternative contemporary ways of harvesting wild rice in India. (A) Beating panicles over a basket, and (B) twisting leaves and stem into bundles that collect shattered grain. These methods, plus swinging a basket over ripening panicles, are used to harvest wild rice in South Asia and West Africa (Oka, 1988).

During domestication, population genetic diversity is reduced as a consequence of selection. Domestication-related genes experience a more severe genetic bottleneck due to selection than neutral genes, as discussed by Doebley et al., (2006) and in this Special Issue by Yamasaki et al. (2007). Estimates of the severity of the genetic bottleneck of domestication based on comparison of genetic diversity found in their wild ancestors vary considerably from about 80 % in maize (Wright and Gaut, 2005), to 40–50 % in sunflower (Liu and Burke, 2006) and as little as 10–20 % in rice (Zhu et al., 2007). Polyploid wheats have suffered two bottlenecks associated with the transition from wild wheat and also due to polyploidy. Thus, hexaploid bread wheat has about 7 % and 30 % of the nucleotide diversity of its D and A/B genome donors, respectively (Dubcovsky and Dvorak, 2007). Determining how much diversity is lost during the genetic bottleneck of domestication can suggest approaches to future crop improvement, such as tapping high diversity gene sources in wild progenitors (Whitt et al., 2002) or transgenic alteration of expression of selected genes (see Yamasaki et al., 2007, in this Special Issue).

Detection of previously undetected domestication-related genes has become possible using QTL analysis and selective sweeps across the genome (Yamasaki et al., 2007). This enables hidden domestication genes to be detected based on the selection profile of comparative sequences. Genomic comparison of crops and their wild progenitors for hidden domestication-related genomic regions is a new approach to detecting potentially useful diversity in wild progenitors for crop improvement.

GENE EVOLUTION

Genome sequencing has enabled the evolution of domestication-related genes to be elucidated. The wild and domesticated wheat species provide an example. The different domesticated wheat cultigens (cultivated species) evolved from hybridization events between wild and cultivated species (to form Triticum aestivum subsp. spelta) or selections from either wild (T. monococcum subsp. monococcum and T. turgidum subsp.dicoccum) or domesticated species (T. aestivum and T. turgidum subsp.durum). In this wild–domesticated polyploid series the grain-hardness locus (Ha) of wheat encodes friabilins that are composed of three proteins. The genes Pina, Pinb and Gsp-1 at the Ha locus encode these three proteins. Two of these genes, Pina and Pinb, were eliminated from both the A and B genomes of wheat after polyploidization into the tetraploid T. turgidum. In the hexaploid, T. aestivum, Pina and Pinb are present coming from the D genome donor, Aegilops tauschii. Comparison of the Ha locus in the same genome in diploid, tetraploid and hexaploidTriticum and Aegilops species revealed numerous genomic rearrangements, such as transposable element insertions, genomic deletions, duplications and inversions (Chantret et al., 2005). Genomic rearrangements at the Ha locus were believed to be mainly caused by illegitimate recombination, where DNA sequences not originally attached to one another become joined, and this type of recombination is considered a major evolutionary mechanism in wheat species (Chantret et al. 2005). The complex evolution of the Ha locus in wheat reflects the remarkably high rate of DNA replacement in wheat genes (Dubcovsky and Dvorak, 2007).

GENE AND GENOME DUPLICATION

An area where DNA technologies have had a particular impact on our understanding of domestication has been in relation to gene and genome duplication. As well as polyploidy, gene duplication is a common evolutionary phenomenon in plants (for review see Moore and Purugganan, 2005). In maize it has been estimated that about a third of genes are tandem duplicates due to unequal recombination or transposition events that have involved gene fragments (Emrich et al., 2007). Rondeau et al. (2005) have shown duplication and subsequent functional specialization of NADH-MDH genes in some, but not all, grasses with C4 photosynthesis. Among duplicated genes are a class called nearlyidentical paralogs (NIPs) that appear to be of recent origin. This class of duplicate gene shares ≥98 % identity. Many NIPs in maize are differentially expressed. This has lead to the suggestion that the variation in this class of duplicate gene provides new variation that may have had a selective advantage during domestication and improvement of maize (Emrich et al., 2007), and again modelling of plant architecture may suggest routes for crop improvement (Guo et al., 2006).

Reflecting the abundance of polyploids in the plant kingdom, many important crops exhibit both allopolyploidy (e.g. wheat, canola, tobacco, peanut and cotton) and autopolyploidy (e.g. watermelon, strawberries, potato and alfalfa). Allopolyploidy results in increased allelic diversity while autopolyploidy results in increased allelic copy number, both of which can lead to novel phenotypes. Since polyploidy is so common in plants they must have some selective advantages. Among the presumed main advantages of polyploidy are fixation of heterosis, duplication enabling evolution of gene function, and alteration of regulation.

The allopolyploid oilseed crop Brassica napus (canola) provides an example of how heterozygosity resulting from polyploidy can affect evolutionarily important traits. Brassica napus is thought to be derived from crosses between B. oleracea (2n = 18, CC genome) and B. rapa (2n = 20, AA genome). Using molecular markers, lines in mapping populations were compared at a transposition site with QTL for seed yield (Osborn et al. 2003a, Quijada et al., 2006, Udall et al., 2006). When the allelic arrangement was similar to that of the parental genotypes, B. oleracea andB. rapa, seed yields were lower. However, when the arrangement of alleles differed from these parental genotypes seed yields were higher. The best explanation for the results of these studies was that intergenomic heterozygosity increased seed yield in B. napus.

In allopolyploid cotton, an ancient polyploidy, it has been shown that some homoeologous genes are assigned to different (sub) functions, with gene expression compartmentalized to different tissue types and gene expression biased between homoeologs (Adams et al., 2003, 2004). Thus, between the genomes of cotton, expression of homoeologous genes is developmentally regulated. It has been suggested that this may provide allopolyploids with greater plasticity in response to stress (Udall and Wendel, 2006). Further understanding of what causes changes in homoeologous gene function may provide avenues to manipulate gene expression.

Gene expression is generally dependent on hierarchically organized networks of regulators. The number of these regulators can be increased several-fold in polyploids and the overall consequences of polyploidy on gene expression at the end of regulatory networks are difficult to predict (Osborn et al., 2003b). In a genome-wide analysis of synthetic allotetraploids between Arabidopsis thaliana and A. arenosa, about 5 % of genes showed divergence from the mid-parent value, suggesting non-additive gene regulation (Wang et al., 2006b). For example, time of flowering in this synthetic allopolyploid was later than both parents. This was found to be the result of the epistatic interactions between two loci, one for flowering from A. thaliana (FLC) and the other from A. arenosa(FRI), that enhances FLC expression and inhibits flowering (Wang et al., 2006a). In hexaploid wheat, latitude of breeding has influenced the selection of genes affecting earliness of flowering, but there is still much genetic diversity relating to both photoperiod and vernalization requirements of the selections (Goldringer et al., 2006). The rapid reprogramming of biological pathways on polyploidization leads to novel variation that may be exploited by plant breeders.

Many breeding programmes involve wide and distant hybridization. These procedures cause dramatic genome change, sometimes leading to unpredictable results. Studies of ancient and modern polyploids provide a means of elucidating the effects of dramatic genome change on gene expression and regulation. Results from such studies should enable breeding programs to achieve the desired results.

ACHIEVING NEW LEVELS OF CROP YIELD AND NEW USES FOR CROPS

The future course of domestication will continue to rely on changes to the architecture, metabolism and physiology of crops. To reach new levels of crop yield and new uses for crops, a combination of applied and also theoretical approaches involving computer or systems' biology models will be required. To cope with new challenges to crop production, efficient approaches to screen germplasm for genes to both biotic and abiotic stresses in wild and cultivated germplasm are needed. Crops primarily used for food are now also being used to meet demands for sustainable fuel supply, while the use of plants for construction timbers and fibres in paper, textiles and composite products continues to increase. To provide value-added crops, new products from crops are being developed. Specialized plant products for processing (such as starches, oils, even plastics) for food, pharmaceutical, cosmetic and industrial uses are required for growing markets (Heslop-Harrison, 2002).

As information accumulates on domestication-related traits and their genome distribution, new avenues to attain higher yield and to tailor-make crops are opened up. Analysis of yield and plant height in a cross between two Japanese rice varieties, 'Koshihikari' and 'Habataki', revealed several QTLs for each trait. One QTL, Gn1a, increased grain productivity and acts by altering the production of the enzyme cytokinin oxidase/dehydrogenase that degrades the phytohormone cytokinin. By reducing the expression of Gn1a, cytokinin accumulates in inflorescence meristems, resulting in an increased number of grains and, hence, a plant with the potential for increased yield (Ashikari et al., 2005). By accumulating a variety of yield-related QTLs for increasing both source to produce photosynthate and sink to accept photosynthate, new levels of yield may be achieved.

Throughout history, plants have been subjected to changing climate, and farmers have adopted new species and varieties to meet the challenges; indeed, post-glacial climate changes may have been one of the factors leading to the origin of agriculture and plant domestication. Climate change is affecting agriculture in the 21st century; some changes will be met within existing adaptations of plants, but other factors such as increased UV and carbon dioxide levels require new selections based on understanding of plant responses. Hidema and Kumagai (2006) reported considerable variation in UVB sensitivity of rice cultivars, which was caused by differences of one or two bases in the CPD (cyclobutane pyrimidine dimer) photolyase, altering the activity of the enzyme. They suggest that the resistance of rice to UVB radiation can therefore be increased by selective breeding or bioengineering of the genes encoding CPD photolyase. Although carbon dioxide enhancement is regularly used to improve glasshouse production, it is not clear how field crops will respond to changes in atmospheric carbon dioxide concentrations, involving complex interactions of phytosythesis with light and dark respiration (Bunce, 2005).

There is no naturally occurring waxy wheat variety but in bread wheat there are waxy loci in each of its three different genomes, A, B and D. The waxy locus encodes starch granule protein 1 (SGP-1). Different isoforms are encoded by genes in each genome (sgpA1, sgpB1 and sgpD1). In the germplasm collection of bread wheat, cultivars lacking one of the three isoforms were found, two cultivars from Korea lacked SGP-A1, one from Japan lacked SPG-B1 and one from Turkey lacked SPG-D1. By making appropriate crosses, these genes were combined in a single plant, resulting in the first waxy wheat, which had a null for all three isoforms of SGP-1 (Yamamori et al., 2000), an example of using markers for identification of alleles and then marker-assisted selection to find the desired allele combination. The effort to produce waxy wheat in Japan has led to its use as an ingredient for improved Japanese-style noodles. These examples show the value of screening germplasm collections with diverse material for useful genetic variation, but also emphasize that it is not always necessary to search in exotic material or employ radical techniques to make innovative progress in plant breeding.

SUPER-DOMESTICATION

We are entering a new era in relation to human understanding of and influence on the genetics of crop domestication. Current research on genes related to crop domestication is providing pieces in the complex jigsaw of gene evolution (see also Hancock, 2005), networking, regulation and expression in our most important plants. Introduction of alien genes through transgenic technology may be difficult (King et al., 2004), but continued advances in crop improvement will depend on understanding the genome and its genes.

Super-domesticates can be constructed with knowledge-led approaches using the range of current technologies. Here, we use the term super-domestication to refer to the processes that lead to a domesticate with dramatically increased yield that could not be selected in natural environments from naturally occurring variation without recourse to new technologies. The array of genome manipulations that have been developed, mainly since the 1980s, enable barriers to gene exchange to be overcome and have lead to super-domesticates with dramatically increased yields, resistances to biotic and abiotic stresses, and with new characters for the marketplace. Hybrid rice (see Cheng et al., 2007, in this Special Issue) can be considered a super-domesticate.

The teams of scientists that support plant breeders are planning and conducting research to change crops radically. For example, changing crops from C3 photosynthesis to C4 photosynthesis is being proposed because it is now known that plants with C3 photosynthesis have enzymes for C4 photosynthesis, and even well-developed C4 pathways can be found at certain locations in C3 plants. In addition, C4-enzyme genes have been inserted into and successfully expressed in rice (Mitchell and Sheehy, 2006). Conversion of a crop from C3 to C4 photosynthesis would certainly be a super-domesticate.

It was with this background of rapid progress being made in studies of crop domestication that a meeting was organized in Tsukuba, Japan, in October 2006, by the National Institute of Agrobiological Sciences (NIAS) and the Organisation for Economic Cooperation and Development (OECD) and supported by Annals of Botany. While the Tsukuba meeting was being planned, a different meeting, entitled Plants, People and Evolution, sponsored by the Linnean Society of London, the Systematics Association and Annals of Botany was in preparation. This meeting was held in London in August 2006. Selected papers from these meetings appear in this Special Issue of Annals of Botany.

Progress in understanding crop domestication, and further advances that lead to greater quantities and improved quality of food crops, depend increasingly on multidisciplinary team approaches (Zeder et al., 2006; Wuchty et al., 2007). Scientists representing a diversity of botanical and crop-science backgrounds, archaeobotanists, crop evolutionary biologists, geneticists, ethnobotanists, plant breeders, statisticians and biotechnology specialists contribute papers in this present volume. The papers include both reviews of topics related to crop domestication and original research articles. Two key papers discuss domestication in the New World (Pickersgill, 2007) and the Old World (Fuller, 2007), while the papers that follow relate to particular crops or groups of crops. Included are papers on crops that have been intensively studied by molecular methods, e.g. maize (Yamasaki et al., 2007), barley (Azhaguvel and Komatsuda, 2007; Pourkheirandish and Komatsuda, 2007), tomato (Bai and Lindhout, 2007), wheat (Waines and Ehdaie, 2007); some whose genomes have been completely sequenced, e.g. rice (Cheng et al., 2007;Sweeney and McCouch, 2007) and sorghum (Dillon et al., 2007), or where sequencing projects are proceeding actively, e.g. soybean (Liu et al., 2007) and common bean (Phaseolus vulgaris; Papa et al., 2007), and some where domestication is still at an early stage, e.g. giant cacti (Casas et al., 2007), artichoke (Sonnante et al., 2007) or banana (Heslop-Harrison and Schwarzacher, 2007). Scientists working on minor crops envy the amount of information being rapidly accumulated on model crops, but by extrapolation information from model species and model crops is already hastening advances in minor crops. This will be particularly true for current genomic initiatives in closely related crops such as the legumes (Weeden, 2007), where data from common bean and soybean will benefit the closely related African and Asian Vigna (Isemura et al., 2007). Knowledge of the sorghum genome can be tapped to make progress in understanding the complex genome of sugarcane (Dillon et al., 2007) and that of the rice genome for banana (Heslop-Harrison and Schwarzacher, 2007). Minor crops, by the very fact that less is known about them, provide the potential of rapidly finding new insights into crop domestication (e.g. Fukunaga et al., 2006).

ACKNOWLEDGEMENTS

The authors would like to thank several anonymous reviewers for help in improving earlier drafts of this paper. The workshop in Tsukuba was sponsored by the Organisation for Economic Co-operation and Development Co-operative Research Programme on Biological Resource Management for Sustainable Agricultural Systems, whose financial support made it possible for most of the invited speakers to participate in the workshop. Funding to pay the Open Access publication charges for this article was also provided by the OECD.

REFERENCES

1. Adams KL, Cronn R, Percifield R, Wendel JF . Genes duplicated by polyploidy show unequal contribution to the transcriptome and organ-specific reciprocal silencing.Proceedings of the National Academy of Sciences USA. 2003.100 p. 4649-4654.
2. Adams KL, Percifield R, Wendel JF . Organ-specific silencing of duplicated genes in a newly synthesized cotton allotetraploid. Genetics 2004;168:2217-2226.
3. Ashikari M, Sakakibara H, Lin S, Yamamoto T, Takashi T, Nishimura A, et al .Cytokinin oxidase regulates rice grain production. Science 2005;309:741-745.
4. Azhaguvel P, Komatsuda T. A phylogenetic analysis based on nucleotide sequence of a marker linked to the brittle rachis locus indicates a diphyletic origin of barley.Annals of Botany 2007;100:1009-1015.

5. Bai Y, Lindhout P. Domestication and breeding of tomatoes: what have we gained and what can we gain in the future? Annals of Botany 2007;100:1085-1094.

6. Bomblies K, Doebley JF . Pleiotropic effects of the duplicate maizeFLORICAULA/LEAFY genes zfl1 and zfl2 on traits under selection during maize domestication. Genetics 2006;172:519-531.

7. Bringhurst RS, Voth V . Breeding octoploid strawberries. Iowa State Journal of Research 1984;58:371-381.

8. Bunce JA . Response of respiration of soybean leaves grown at ambient and elevated carbon dioxide concentrations to day-to-day variation in light and temperature under field conditions. Annals of Botany 2005;95:1059-1066.

9. Cao Q, Lu B-R, Xia H, Rong J, Sala F, Spada A, Grassi F . Genetic diversity and origin of weedy rice (Oryza sativa f. spontanea) populations found in north-eastern China revealed by simple sequence repeat (SSR) markers. Annals of Botany 2006;98:1241-1252.

10. Casas A, Otero-Arnaiz A, Pérez-Negrón E, Valiente-Banuet A . In situ management and domestication of plants in Mesoamerica. Annals of Botany 2007;100:1101-1115.

11. Chantret N, Salse J, Sabot F, Rahman S, Bellec A, Laubin B, Dubois I, Dossat C, et al .Molecular basis of evolutionary events that shaped the Hardness locus in diploid and polyploid wheat species (Triticum and Aegilops). The Plant Cell 2005;17:1033-1045.

12. Cheng S-H, Zhuang J-Y, Fan Y-Y, Du J-H, Cao L-Y . Progress in research and development in hybrid rice: a super-domesticate in China. Annals of Botany2007;100:959-966.

13. Clark RM, Linton E, Messing J, Doebley JF. Pattern of diversity in the genomic region near the maize domestication gene tb1. Proceedings of the National Academy of Sciences USA. 2004.101 p. 700-707.

14. Clark RM, Wagler TN, Quijada P, Doebley J. A distant upstream enhancer at the maize domestication gene tb1 has pleiotropic effects on plant and inflorescent architecture. Nature Genetics 2006;38:594-597.

15. Consonni G, Gavazzi G, Dolfini S. Genetic analysis as a tool to investigate the molecular mechanisms underlying seed development in maize. Annals of Botany2005;96:353-362.

16. Crawley MJ, Brown SL. Seed limitation and the dynamics of feral oilseed rape on the M25 motorway. Proceedings of the Royal Society of London Series B. 1995.259 p. 49-54.

17. Darwin C. The origin of species by natural selection or the preservation

of favoured races in the struggle for life. 1st edn. London: John Murray; 1859.

18. Dillon SL, Shapter FM, Henry RJ, Cordeiro G, Izquierdo L, Lee SL. Domestication to crop improvement: genetic resources for Sorghum and Saccharum (Andropogoneae).Annals of Botany 2007;100:975-989.
19. Doebley J, Lukens L. Transcriptional regulators and the evolution of plant form. The Plant Cell 1998;10:1075-1082.
20. Doebley J, Gaut BS, Smith BD. The molecular genetics of crop domestication. Cell2006;127:1309-1321.
21. Doust A. Architectural evolution and its implications for domestication in grasses.Annals of Botany 2007;100:941-950.
22. Dubcovsky J, Dvorak J. Genome plasticity: a key factor in the success of polyploid wheat under domestication. Science 2007;316:1862-1866.
23. Emrich SJ, Li L, Wen T-J, Yandeau-Nelson MD, Fu Y, Guo L, et al. Nearly identical paralogs: implications for maize (Zea mays L.) genome evolution. Genetics2007;175:429-439.
24. Fukunaga K, Ichitani K, Kawase M. Phylogenetic analysis of the rDNA intergenic spacer subrepeats and its implications for the domestication history of foxtail millets.Setaria italica. Theoretical and Applied Genetics 2006;113:261-269.
25. Fuller DQ. Contrasting patterns in crop domestication and domestication rates: recent archaeobotanical insights from the Old World. Annals of Botany 2007;100:903-924.
26. Goldringer I, Prouin C, Rousset M, Galic N, Bonnin I. Rapid differentiation of experimental populations of wheat for heading time in response to local climatic conditions. Annals of Botany 2006;98:805-817.
27. Guo Y, Ma Y, Zhan Z, Li B, Dingkuhn M, Luquet D, De Reffye P. Parameter optimization and field validation of the functional-structural model GREENLAB for maize.Annals of Botany 2006;97:217-230.
28. Hajjar R, Hodgkin T. The use of wild relatives in crop improvement: a survey of developments over the last 20 years. Euphytica 2007;156:1-13.
29. Hammer K. Das Domestikationssyndrom. Kulturpflanze 1984;32:11-34.
30. Hancock JF. Contributions of domesticated plant studies to our understanding of plant evolution. Annals of Botany 2005;96:953-963.
31. Heslop-Harrison JS. Exploiting novel germplasm. Australian Journal of Agricultural Research 2002;53:1-7.

32. Heslop-Harrison JS, Schwarzacher T. Domestication, genomics and the future for banana. Annals of Botany 2007;100:1073-1084.
33. Hidema J, Kumagai T. Sensitivity of rice to ultraviolet-B radiation. Annals of Botany2006;97:933-942.
34. Hubbard L, McSteen P, Doebley J, Hake S. Expression patterns and mutant phenotype of teosinte branched1 correlate with growth suppression in maize and teosinte. Genetics 2002;162:1927-1935.
35. Isemura T, Kaga A, Konishi S, Ando T, Tomooka N, Han OK, Vaughan DA. Genome dissection of traits related to domestication in azuki bean (Vigna angularis) and comparison with other warm-season legumes. Annals of Botany 2007;100:1053-1071.
36. King D, Dalton H, Heslop-Harrison JS. GM Science Review. Second Report. An open review of the science relevant to GM crops and food based on interests and concerns of the public. London: HMSO; 2004 [Accessed June 2007].http://www.gmsciencedebate.org.uk/report/pdf/gmsci-report2-full.pdf.
37. Komatsuda T, Pourkheirandish M, He C, Azhaguvel P, Kanamori H, Perovic D, et al.Six-rowed barley originated from a mutation in a homeodomain-leucine zipper 1-class homeobox gene. Proceedings of the National Academy of Sciences USA. 2007.104 p.1424-1429.
38. Konishi S, Izawa T, Lin S-Y, Ebana K, Fukuta Y, Sasaki T, Yano M. An SNP caused loss of seed shattering during rice domestication. Science 2006;312:1392-1396.
39. Li C, Zhou A, Sang T. Rice domestication by reduced shattering. Science2006;311:1936-1939.
40. Lin Z, Griffith ME, Li X, Zhu Z, Tan L, Fu Y, Zhang W, Wang X, Xie D, Sun C. Origin of seed shattering in rice (Oryza sativa L.). Planta 2007;226:11-20.
41. Liu A, Burke JM. Patterns of nucleotide diversity in wild and cultivated sunflowers.Genetics 2006;173:321-330.
42. Liu B, Fujita T, Yan Z-H, Sakamoto S, Xu D, Abe J. QTL mapping of domestication related traits in soybean (Glycine max). Annals of Botany 2007;100:1027-1038.
43. Mitchell PL, Sheehy JE. Supercharging rice photosynthesis to increase yield. New Phytologist 2006;171:688-693.
44. Moore RC, Purugganan MD. The evolutionary dynamics of plant duplicate genes.Current Opinion in Plant Biology 2005;8:122-128.
45. Morrell PL, Clegg MT. Genetic evidence for a second domestication of barley (Hordeum vulgare) east of the Fertile Crescent. Proceedings of the National Academy of Sciences USA. 2007.104 p. 3289-3294.

46. Nesbitt TC, Tanksley SD. Comparative sequencing in the genus Lycopersicon:implications for the evolution of fruit size in the domestication of cultivated tomatoes.Genetics 2002;162:365-379.

47. Oka HI. Origin of cultivated rice. Amsterdam: Elsevier; 1988.

48. Osborn TC, Butrulle DV, Sharpe AG, Pickering KJ, Parkin IAP, Parker JS, Lydiate DJ.Detection and effects of a homeologous reciprocal transpostion in Brassica napus. Genetics. 2003;165:1569-1577.

49. Osborn TC, Pires JC, Birchler JA, Auger DL, Chen ZJ, Lee HS, et al. Understanding mechanisms of novel gene expression in polyploids. Trends in Genetics 2003;19:141-147.

50. Østergaard L, Kempin SA, Bies D, Klee HJ, Yanofsky MF. Pod shatter-resistantBrassica fruit produced by ecotopic expression of the FRUITFULL gene. Plant Biotechnology Journal 2006;4:45-51.

51. Palaisa K, Morgante M, Tingey S, Rafalski A. Long-range patterns of diversity and linkage disequilibrium surrounding the maize Y1 gene are indicative of an asymmetric selective sweep. Proceedings of the National Academy of Sciences USA. 2004.101 p.9885-9890.

52. Papa R, Bellucci E, Rossi M, Leonardi S, Rau D, Gepts P, Nanni L, Attene G. Tagging the signatures of domestication in common bean (Phaseolus vulgaris) by means of pooled DNA samples. Annals of Botany 2007;100:1039-1051.

53. Paterson AH. What has QTL mapping taught us about plant domestication? New Phytologist 2002;154:591-608.

54. Phillips RL. Genetic tools from nature and the nature of genetic tools. Crop Science2006;46:2245-2252.

55. Pickersgill B. Domestication of plants in the Americas: insights from Mendelian and molecular genetics. Annals of Botany 2007;100:925-940.

56. Pourkheirandish M, Komatsuda T. The importance of barley genetics and domestication in a global perspective. Annals of Botany 2007;100:999-1008.

57. Purugganan MD, Boyles AL, Suddith JI. Variation and selection at the CAULIFLOWERfloral homeotic gene accompanying the evolution of domesticated Brassica oleracea.Genetics 2000;155:855-862.

58. Quijada PA, Udall JA, Lambert B, Osborn TC. Quantitative trait analysis of seed yield and other complex traits in hybrid spring rapeseed (Brassica napus L.). 1. Identification of genomic regions from winter germplasm. Theoretical and Applied Genetics2006;113:549-561.

59. Rondeau P, Rouch C, Besnard G. NADP-malate dehydrogenase gene evolution in Andropogoneae (Poaceae): gene duplication followed by

sub-functionalization. Annals of Botany 2005;96:1307-1314.

60. Sang T, Ge S. The puzzle of rice domestication. Journal of Integrative Plant Biology2007;49:760-768.

61. Simons KJ, Fellers JP, Trick HN, Zhang Z, Tai Y-S, Gill BS, Faris JD. Molecular characterization of the major wheat domestication gene Q. Genetics. 2006;172:547-555.

62. Sonnante G, Pignone D, Hammer K. The domestication of artichoke and cardoon: from Roman times to the genomic age. Annals of Botany 2007;100:1095-1100.

63. Sweeney M, McCouch S. The complex history of the domestication of rice. Annals of Botany 2007;100:951-957.

64. Takeda T, Suwa Y, Suzuki M, Kitano H, Ueguchi-Tanaka M, Ashikari M, Matsuoka M, Ueguchi C. The OsTB1 gene negatively regulates lateral branching in rice. The Plant Journal 2003;33:513-520.

65. Tanksley SD, McCouch SR. Seed banks and molecular maps: unlocking genetic potential from the wild. Science 1997;277:1063-1066.

66. Tanno K, Wilcox G. How fast was wild wheat domesticated? Science2006;311:1886.

67. Udall JA, Wendel JF. Polyploidy and crop improvement. Crop Science2006;46(S1):3-14.

68. Udall JA, Quijada PA, Lambert B, Osborn TC. Quantitative trait analysis of seed yield and other complex traits in hybrid spring rapeseed (Brassica napus L.). 2. Identification of alleles from unadapted germplasm. Theoretical and Applied Genetics2006;113:597-609.

69. Waines JG, Ehdaie B. Domestication and crop physiology: roots of green-revolution wheat. Annals of Botany 2007;100:991-998.

70. Wang J, Tian L. Lee HS, Chen ZJ. Nonadditive regulation of FRI and FRC loci meditates flowering-time variation in Arabidopsis allopolyploids. Genetics 2006;173:965-974.

71. Wang J, Tian L. Lee HS, Wei NE, Jiang H, Watson B, et al. Genomewide nonadditive gene regulation in Arabidopsis allotetraploids. Genetics 2006;172:507-517.

72. Wang R-L, Stec A, Hey J, Lukens L, Doebley J. The limits of selection during maize domestication. Nature 1999;398:236-239.

73. Weeden NF. Genetic changes accompanying the domestication of Pisum sativum: is there a common genetic basis to the. Annals of Botany 2007;100:1017-1025.'domestication syndrome' for legumes?

74. Whitt SR, Wilson LM, Tenaillon MI, Gaut BS, Buckler IV ES. Genetic diversity and selection in the maize starch pathway. Proceedings of

the National Academy of Sciences USA. 2002.99 p. 12959-12962.

75. Wright SI, Gaut BS. Molecular population genetics and the search for adaptive evolution in plants. Molecular Biology and Evolution 2005;22:506-519.

76. Wuchty S, Jones BJ, Uzzi B. The increasing dominance of teams in production of knowledge. Science 2007;316:1036-1039.

77. Yamamori M, Fujita S, Hayakawa K, Matsuki J, Yasui T. Genetic elimination of a starch granule protein, SGP-1, of wheat generates an altered starch with apparent high amylose. Theoretical and Applied Genetics 2000;101:21-29.

78. Yamasaki M, Wright SI, McMullen MD. Genomic screening for artificial selection during domestication and improvement in maize. Annals of Botany 2007;100:967-973.

79. Zeder MA, Bradley DG, Emshwiller E, Smith BD. Documenting domestication: new genetic and archaeological paradigms. Berkeley, CA: University of California Press;2006.

80. Zhang ZH, Qu X, Wan S, Chen Sh, Zhu Y. Comparison of QTL controlling seedling vigour under different temperature conditions using recombinant inbred lines in rice (Oryza sativa). Annals of Botany 2005;95:423-429.

81. Zhu Q, Zheng X, Luo J, Gaut BS, Ge S. Multilocus analysis of nucleotide variation ofOryza sativa and its wild relatives: severe bottleneck during domestication of rice.Molecular Biology and Evolution 2007;24:875-888.

Chapter 10

A DUAL GENE-SILENCING VECTOR SYSTEM FOR MONOCOT AND DICOT PLANTS

Ming-Ru Liou[1,2], Ying-Wen Huang[1], Chung-Chi Hu[1], Na-Sheng Lin[1,2,*] andYau-Heiu Hsu[1,*]

[1]Graduate Institute of Biotechnology, National Chung Hsing University, Taichung, Taiwan, [2]Institute of Plant and Microbial Biology, Academia Sinica, Taipei, Taiwan

SUMMARY

Plant virus-based gene-silencing vectors have been extensively and successfully used to elucidate functional genomics in plants. However, only limited virus-induced gene-silencing (VIGS) vectors can be used in both monocot and dicot plants. Here, we established a dual gene-silencing vector system based on *Bamboo mosaic virus* (BaMV) and its satellite RNA (satBaMV). Both BaMV and satBaMV vectors could effectively silence endogenous genes in *Nicotiana benthamiana* and *Brachypodium distachyon*. The satBaMV vector could also silence the green fluorescent protein (GFP) transgene in *GFP* transgenic *N. benthamiana*. *GFP* transgenic plants co-agro-inoculated with BaMV and satBaMV vectors carrying sulphur and *GFP* genes, respectively, could simultaneously silence both genes. Moreover, the silenced plants could still survive with the silencing of genes essential for plant development such as *heat-shock protein 90 (Hsp90)* and *Hsp70*. In addition, the satBaMV- but not BaMV-based vector could enhance gene-silencing efficiency in newly emerging leaves of *N. benthamiana* deficient in RNA-dependant RNA polymerase 6. The dual gene-silencing vector system of BaMV and satBaMV provides a novel tool for comparative functional studies in monocot and dicot plants.

INTRODUCTION

Gene silencing is the major antiviral defence mechanism in plants. During plant virus infection, double-stranded RNA (dsRNA) intermediates are formed and cleaved into small interfering RNAs (siRNAs) by Dicer-like (DCL) proteins. These siRNAs, referred to as primary siRNAs, subsequently incorporate into the RNA-induced silencing complex (RISC), triggering the surveillance mechanism for the degradation or translation arrest of their cognate RNAs.

Gene silencing is amplified by plant-encoded RNA-dependent RNA polymerase (RDR); RDR6 uses primary siRNAs as a primer to anneal viral RNA for dsRNA synthesis; the siRNAs are then processed into secondary siRNAs by DCL proteins. RDR6 protein is important for the biosynthesis of secondary siRNAs and systemic silencing in the antiviral defence response (Garcia-Ruiz *et al.*, 2010; Wang *et al.*, 2010). Recently, the role of RDR6 in limiting virus invasion into newly emerging leaves was demonstrated with RDR6-deficient (*rdr6*) *Nicotiana benthamiana*. Knocking down *RDR6* in *N. benthamiana* caused hypersusceptibility to *Hop stunt viroid* (HSVd) (Gómez *et al.*, 2008), *Potato spindle tuber viroid* (PSTVd) (Di Serio *et al.*, 2010), *Potato virus X* (PVX) (Qu *et al.*, 2005; Schwach *et al.*, 2005), *Potato virus Y* (PVY) and Y satellite of *Cucumber mosaic virus* (CMV) (Schwach *et al.*, 2005). However, the efficiency of gene silencing induced by PVX and *Plum pox virus* (PPV) and RNA-directed DNA methylation was reduced in *rdr6 N. benthamiana* (Vaistij and Jones, 2009). The decreased efficiency was related to the accumulation of viral siRNA, so RDR6 may have roles in mediating the generation of viral primary and secondary siRNA for surveillance virus infection (Schwach *et al.*, 2005; Vaistij and Jones, 2009; Vaistij *et al.*, 2002).

With plant virus-based vector engineering of a gene fragment from a plant, the inoculated plant could trigger a sequence-specific RNA degradation process, referred to as virus-induced gene silencing (VIGS) (Baulcombe, 1999). VIGS has been used extensively for more than a decade to characterize gene functions in plants. It has the advantages of rapidness, efficiency, and specificity and could also be used in high-throughput functional genomics in plants (Dong *et al.*, 2007; Lu *et al.*, 2003; Pacak *et al.*, 2010a; Yuan *et al.*, 2011). Many RNA and DNA viruses have been modified to serve as gene-silencing vectors (Burch-Smith *et al.*, 2004; Godge *et al.*, 2008; Senthil-Kumar and Mysore, 2011; Unver and Budak, 2009). The commonly used vectors are *Barley strip mosaic virus* (BSMV) (Holzberg *et al.*, 2002), PVX (Ruiz *et al.*, 1998) and *Tobacco rattle virus* (TRV) (Liu *et al.*, 2002; Ratcliff *et al.*, 2001; Valentine *et al.*, 2004). Subviral agents developed as VIGS vectors include *Satellite tobacco mosaic virus* (STMV) (Gosselé *et al.*, 2002), *Tobacco curly shoot virus* DNA1 (TbCSV DNA1) (Huang*et al.*, 2009), *Tomato bushy stunt virus* (TBSV) defective interfering RNAs (TBSV DIs) (Hou and Qiu, 2003) and *Tomato yellow leaf curl China virus* DNAβ (TYLCCNV DNAβ) (Tao and Zhou, 2004). However, the use of satellite RNA (satRNA) as a vector to induce gene silencing has not yet been reported.

SatRNAs are subviral agents exclusively associated with many groups of plant viruses. They share little or no sequence homology with the helper virus genome but depend on their helper viruses for replication and encapsidation (Hu *et al.*, 2009; Huang *et al.*, 2010; Roossinck*et al.*, 1992; Simon *et al.*, 2004). Of the known single-stranded satRNAs (ss-satRNAs), most with genomes <0.7 kb do not translate any detectable proteins *in vivo* and allow insertion of a limited size of foreign genes into the genome (Roossinck *et al.*, 1992; Simon *et al.*,2004). However, few ss-satRNAs with genomes >0.7 kb can encode a nonstructural protein (Hu *et al.*, 2009), but most of these proteins are required for the replication of ss-satRNAs (Hemmer *et al.*, 1993; Hu *et al.*, 2009; Liu and Cooper, 1993). P20, encoded by the satRNA of *Bamboo mosaic virus* (satBaMV), is the only exception so far and is not essential for satBaMV replication (Lin *et al.*, 1996).

SatBaMV is a linear RNA molecule of 836 nt that contains an open reading frame (ORF) for the 20-kDa (P20) flanked by a 5′ untranslated region (UTR) of 159 nt and a 3′ UTR of 129 nt (Lin and Hsu, 1994). P20 is an RNA-binding protein that can be replaced with other proteins such as chloramphenicol acetyltransferase (CAT). Thus, satBaMV has been developed as a useful plant expression vector (Lee *et al.*,2000; Lin *et al.*, 1996). The helper virus BaMV, a member of the *Potexvirus* genus, contains a single-stranded positive-sense RNA genome with 5 conserved ORFs (Lin *et al.*, 1994; Yang *et al.*, 1997). BaMV has been developed as a vaccine vector for expressing the epitopes of foot-and-mouth disease virus (Yang *et al.*, 2007) and infectious bursal disease virus (Chen *et al.*, 2012).

In this study, we established a dual gene-silencing vector system using BaMV and satBaMV for functional study in monocot and dicot plants. The satBaMV vector could efficiently silence endogenous and transformed genes and together with the BaMV vector could silence 2 genes simultaneously. In addition, silencing was enhanced with the satBaMV vector but compromised with the BaMV vector in *rdr6 N. benthamiana*. To the best of our knowledge, this is the first development of a satRNA-based vector system for VIGS application in monocots and dicots.

RESULTS

Development of satBaMV as a Gene-Silencing Vector

Previously, we showed that P20 of satBaMV was not essential for satBaMV replication and could be replaced with a reporter gene (Lin*et al.*, 1996). Therefore, we replaced the P20 sequence of satBaMV with a fragment of sulphur (*Sul*) gene, a component of the magnesium chelatase complex required for chlorophyll production (Koncz *et al.*, 1990), to generate pCF4:Sul-C2 (Figure 1a). In addition, we fused different sizes of P20 gene with *Sul* gene to generate the following constructs: pCF4:Sul-C1, pCF4:Sul-N1, pCF4:Sul-N2, pCF4:Sul-N3, pCF4:Sul-N4 and pCF4:Sul-N5 (Figure 1a). *N. benthamiana* plants were mechanically inoculated with plasmid pCB, the infectious cDNA clone of BaMV (Lin *et al.*, 2004), alone or with pCF4 or derived constructs. At 14 days postinoculation (dpi), all inoculated plants, except those co-inoculated with pCF4:Sul-N5 (Figure 1b, 10), started to show the yellow and white colour phenotype of *Sul* silencing in newly developed leaves (Figure 1b, 4–9). Co-inoculation with pCF4:Sul-N3 and pCF4:Sul-N4 produced a clearer and more extensive silencing phenotype than with other constructs (Figure 1b, 8–9).

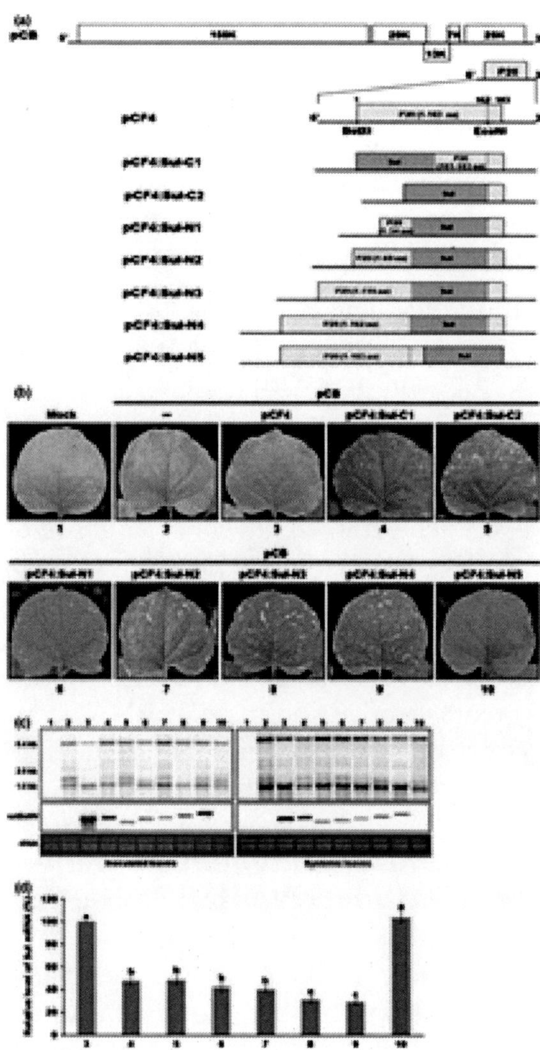

Figure 1. SatBaMV-based vector used to silence *Sul* gene in *Nicotiana benthamiana*. (a) Genome organization of BaMV, satBaMV expression vectors and schematic representations of constructs of satBaMV vectors containing *Sul*. (b) Phenotypes of *N. benthamiana* undergoing VIGS of endogenous *Sul* gene at 14 dpi. (c) Northern blot hybridization of BaMV and satBaMV RNA levels in *N. benthamiana* inoculated with BaMV alone or co-inoculated with satBaMV or chi-

meric satBaMV. Blots were hybridized with DIG-labelled RNA complementary to the 3′ UTR of BaMV (Lin *et al.*, 1993) or 3′ UTR of satBaMV RNA (Lin *et al.*, 1996; Vijayapalani *et al.*, 2012). Positions of BaMV genomic RNA (6.4 kb), 2 subgenomic RNA (2.0 and 1.0 kb) and satBaMV RNA are indicated on the left side of the panel. Ethidium bromide staining of the gel before blotting shows approximately equal loading, as revealed by ribosomal RNA (rRNA) abundance in each lane (bottom panel). (d) Q-RT-PCR analysis of relative mRNA level of *Sul* in WT and silenced*N. benthamiana*. Data are mean ± SD relative mRNA level of *Sul* compared to pCF4 (lane 3) control from 3 independent experiments. The mRNA level of*Sul* was normalized to that of actin. The value of the vector control was set to 100. Letters indicate significant differences by Scheffe's S test at $P = 0.05$. The lane numbers are as in Figure 1b.

The infected plants also developed viral symptoms, but this did not affect the silencing phenotype at the early stage of infection. Control plants co-inoculated with pCB and pCF4 vector without the *Sul* gene (empty vector) remained green (Figure 1b, 3). Northern blot analysis revealed that the levels of satBaMV and its derivatives, except pCF4:Sul-N5, accumulated substantially in inoculated and systemic leaves (Figure 1c). The mRNA levels of cognate *Sul* were reduced about 52%, 52%, 58%, 60%, 69% and 71% in leaves co-inoculated with pCF4:Sul-C1, pCF4:Sul-C2, pCF4:Sul-N1, pCF4:Sul-N2, pCF4:Sul-N3 and pCF4:Sul-N4, respectively (Figure 1d). The silencing phenotype was well associated with *Sul* mRNA levels. Because pCF4:Sul-N4 triggered the greatest reduction in *Sul* mRNA accumulation among all satBaMV constructs in three independent biological replicates, we used this vector for future constructions.

Silencing of Essential Endogenous Genes in *n. Benthamiana* with Satbamv Vector

Besides using mechanical inoculation, we also used *Agrobacterium* infiltration (agro-infiltration), the most common method in VIGS (Vaghchhipawala *et al.*, 2010), to silence endogenous genes. Again, we targeted 2 visual marker genes, such as tobacco *Sul* and phytoene desaturase (*PDS*), required for carotenoid production against photobleaching (Kumagai *et al.*, 1995). To target *Sul*, we co-agro-infiltrated chimeric satBaMV (pKF4:NbSul) with pKB into *N. benthamiana* (Figure 2a and b, 1–3). The infiltrated plants started to show a yellow-white colour along the veins on upper uninfiltrated leaves at 7 dpi (Figure 2b, 4–7 and c, 4). The newly developed leaves of infected plants showed highly uniform yellowing, a typical knockout phenotype of *Sul* inhibition, so *Sul* had been silenced by pKF4:NbSul (Figure 2c, 4).

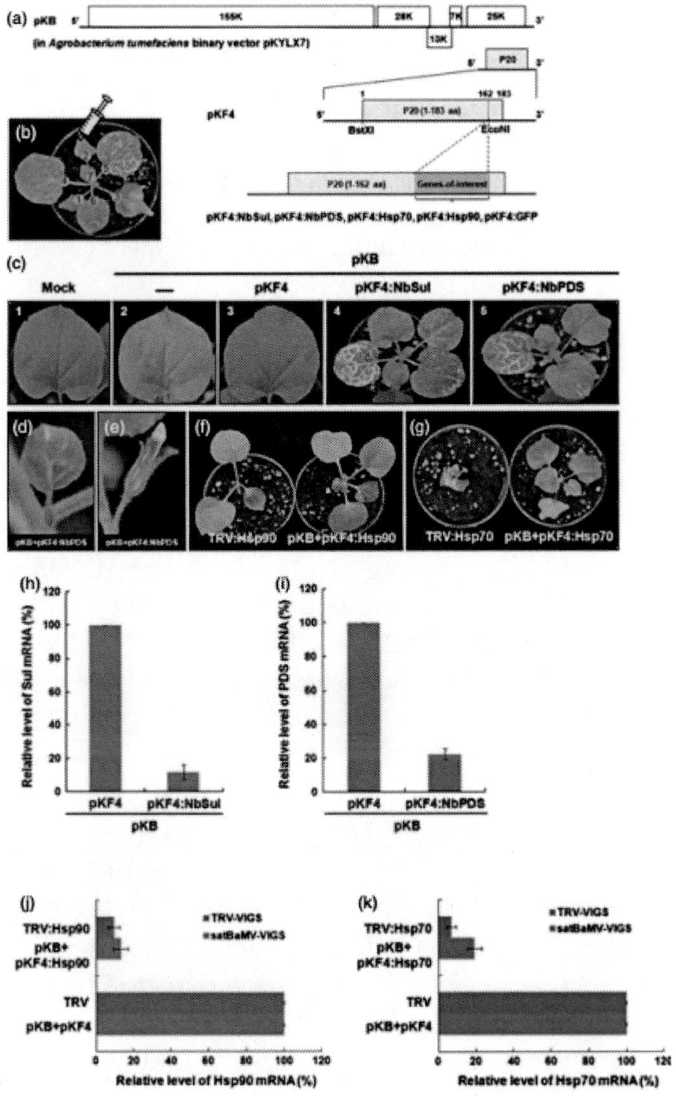

Figure 2. Silencing of endogenous genes in *Nicotiana benthamiana* with a satBaMV-based vector. (a) Genome organization of BaMV, satBaMV RNA and satBaMV-derived constructs for VIGS. (b) The *N. benthamiana* leaves were co-

infiltrated with pKB and pKF4:NbSul (1-3) and induced *Sul* gene silencing in fourth to seventh leaves (4–7) at 14 dpi. (c) Phenotypes of plants agro-infiltrated with a binary vector KYLX7 only (Mock) (1) or pKB (2) or co-infiltrated with pKF4 (3), pKF4:NbSul (4) or pKF4:NbPDS (5) at 14 dpi. (d–e) SatBaMV-induced phytoene desaturase (*PDS*) gene silencing caused photobleaching in axillary shoots (d) and sepals (e) at 35 dpi. (f) Silencing *Hsp90* disrupted leaf development, which resulted in severe crinkling and stunted growth at 21 dpi with TRV- and satBaMV-based VIGS. (g) Silencing *Hsp70* caused death with TRV-based VIGS, but satBaMV-based VIGS produced defective leaves and stunted growth at 21 dpi. (h–i) Q-RT-PCR analysis of relative mRNA levels of *Sul* (h) and *PDS* (i) in WT and silenced *N. benthamiana*. (j–k) Q-RT-PCR analysis of relative mRNA levels of *Hsp90* (j) and *Hsp70* (k) in silenced *N. benthamiana* at 16 dpi. Data are mean ± SD from 3 independent experiments.

The mRNA levels of *Sul* were markedly reduced in silenced leaves (Figure 2h). This co-infiltration of *N. benthamiana* plants resulted in 100% infection in all independent experiments, and all plants developed the same phenotypes. The plants with mock silencing, infiltrated with pKB or co-infiltrated with pKF4, showed mosaic symptoms but no change in leaf colour (Figure 2c, 1–3).

To target *PDS*, we co-infiltrated pKB and pKF4:NbPDS in *N. benthamiana* plants (Figure 2a), which produced a white photobleaching phenotype along leaf veins that was typical of *PDS* inhibition at 7 dpi (Figure 2c, 5). The mRNA level of *PDS* was reduced by almost 80% in silenced leaves as compared to plants co-infiltrated with pKB and pKF4 (Figure 2i). In addition to the phenotype around veins, axillary shoots and sepals of *N. benthamiana* showed the silencing phenotype (Figure 2c, 5; d and e).

TRV is the most widely used VIGS vector for dicots because of its effective silencing of endogenous genes throughout the entire plant (Liu*et al.*, 2002; Ratcliff *et al.*, 2001; Valentine *et al.*, 2004). However, TRV may affect vitality when silencing development-related genes, such as *heat-shock protein 90* (*Hsp90*) and *Hsp70* (Liu *et al.*, 2004). To determine whether the satBaMV vector could overcome this drawback, we compared silencing of *Hsp90* and *Hsp70* genes with the TRV and satBaMV vector. Silencing *Hsp90* with TRV:Hsp90 disrupted leaf development, as shown by severe crinkling and stunting at 21 dpi (Figure 2f). Similarly, silencing *Hsp70* with TRV:Hsp70 caused plant death (Figure 2g). In contrast, satBaMV-induced VIGS produced leaves only slightly defective and stunted at 21 dpi (Figure 2f and g). TRV:Hsp90 and TRV:Hsp70 silencing was stronger, at 90.5% and 92.9%, respectively, than satBaMV-based pKB+pKF4:Hsp90 and pKB+pKF4:Hsp70 silencing, at 86.8% and 80.7%, respectively (Figure 2a, j and k). Therefore, plant growth may be greater with satBaMV than TRV silencing. Although both of these vector systems could significantly induce silencing, the satBaMV vector may be suitable for silencing essential endogenous genes required for plant growth and development.

Enhanced satBaMV-induced gene silencing in *rdr6 N. benthamiana*

To examine the possible role of RDR6 in satBaMV-VIGS, we agro-infiltrated wild type (WT) and *rdr6 N. benthamiana* (Schwach *et al.*, 2005) with pKB alone or with pKF4. At 16 dpi, all infiltrated *rdr6* plants showed mosaic symptoms in newly emerged young leaves (Figure 3a, 7–8), whereas WT plants showed mosaic symptoms only in mature leaves but not in newly emerged young leaves (Figure 3a, 2–3). Northern blot analysis confirmed that the accumulation of BaMV and satBaMV RNA well matched the symptom severity (Figure 3b–d). The difference in BaMV and satBaMV RNA accumulation was only slight in inoculated and 6th systemic leaves of WT and *rdr6* plants (Figure 3b–c, lanes 2–3 vs. lanes 7–8). However, BaMV and satBaMV RNA levels were greater in newly emerged leaves of *rdr6* plants. This difference was more visible and greater for the topmost (8th) systemic leaves of *rdr6* plants at 16 dpi (Figure 3d, lanes 2–3 vs. lanes 7–8). Reduced expression of NbRDR6 may result in hypersusceptibility to BaMV and satBaMV infection in newly emerged leaves.

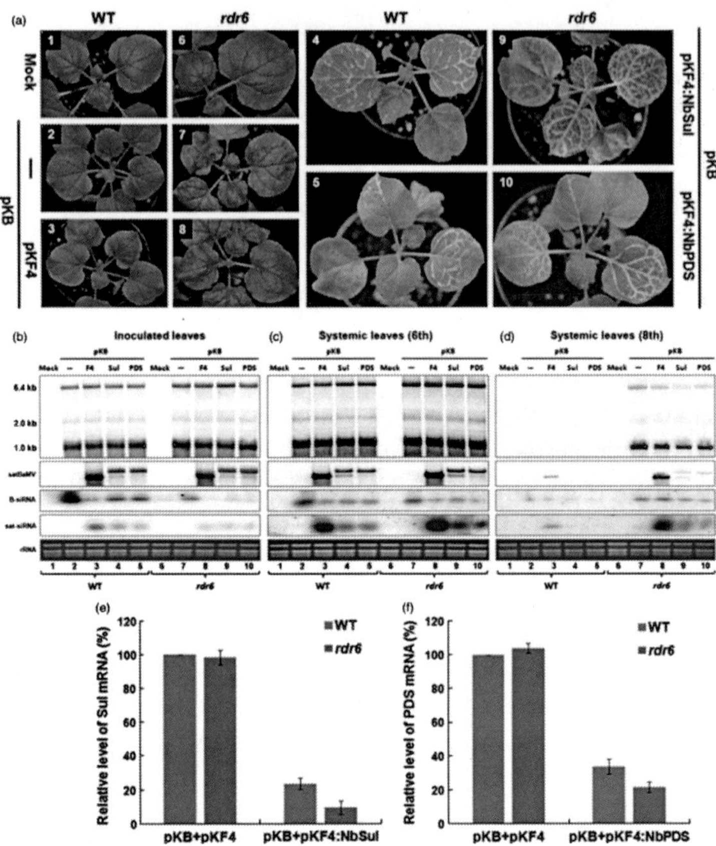

Figure 3. Enhanced satBaMV-induced gene silencing in *rdr6 Nicotiana benthamiana*. (a) WT and *rdr6* plants agro-infiltrated with pKB or co-infiltrated with pKF4 or its derived constructs to silence *Sul* and *PDS* genes at 16 dpi. (b–d) Northern blot analyses of RNA levels of BaMV, satBaMV and siRNA in inoculated (b), 6th systemic (c) and 8th systemic (d) leaves of WT and *rdr6*plants at 16 dpi. Lane numbers are the same as in Figure 3a. (e–f) Q-RT-PCR analysis of relative mRNA levels of *Sul* (e) and *PDS* (f) in silenced 6th systemic leaves of WT and *rdr6* plants at 16 dpi. Vector control value was set to 100%. Data are mean ± SD from 3 independent experiments.

Because BaMV and satBaMV hyperaccumulated in *rdr6 N. benthamiana* plants, we next tested whether this hyperaccumulation affected the efficiency of satBaMV-VIGS. WT and *rdr6* plants were infiltrated with pKB alone or co-infiltrated with pKF4:NbSul, pKF4:NbPDS or pKF4. At 16 dpi, WT and *rdr6* plants co-infiltrated with pKB and pKF4:NbSul or pKF4:NbPDS showed a yellow-white phenotype as compared with mock inoculation (Figure 3a, 4–5 vs. 1; 9–10 vs. 6). Moreover, the yellow-white colour spreads into the uninfiltrated topmost leaves of *rdr6*with pKB+pKF4:NbSul and pKB+pKF4:NbPDS infection (Figure 3a, 9–10), whereas the uppermost leaves were largely free of the *Sul* and*PDS* silencing phenotype in WT plants (Figure 3a, 4–5). Northern blot analysis with BaMV- and satBaMV-specific probes confirmed that the RNA levels of BaMV, satBaMV and chimeric satBaMV, respectively, were slightly higher in inoculated and 6th systemic leaves of *rdr6* than WT plants (Figure 3b–c, lanes 9–10 vs. lanes 4–5). Of note, the expression of pKF4, pKF4:NbSul and pKF4:NbPDS was barely detectable with the satBaMV-specific probe in the 8th systemic leaves of WT plants (Figure 3d, lanes 3–5) but was substantially accumulated with higher levels in *rdr6* plants than in WT plants (Figure 3d, lanes 8–10).

To further detect the accumulation of siRNA, total RNA was separated in denaturing RNA gel and blots were hybridizated with BaMV- and satBaMV-specific probes. The accumulation of BaMV and satBaMV siRNA was significantly lower in inoculated leaves of *rdr6* plants than in WT leaves (Figure 3b, lanes 2–5 vs. lanes 7–10). Similarly, the level of BaMV siRNA was lower in the 6th systemic leaves of *rdr6* plants than WT leaves, but the level of satBaMV siRNA was significantly higher than in WT leaves (Figure 3c, lanes 2–5 vs. lanes 7–10). Nevertheless, the levels of these two siRNAs were markedly higher in 8th systemic leaves of *rdr6* plants than WT plants (Figure 3d, lanes 2–5 vs. lanes 7–10). Thus, RNAs and siRNAs of BaMV and satBaMV may have hyperaccumulated in newly emerged leaves of *rdr6* plants. These silencing phenotypes well matched a lower mRNA accumulation of *Sul* and *PDS* in *rdr6* than WT leaves (Figure 3e and f).

Previously, RDR6 was found required for sense-transgene silencing (Dalmay *et al.*, 2000). Using agro-infiltration with T-DNA constructs could trigger RDR6-dependent transgene silencing (Dunoyer *et al.*, 2006). To eliminate this possibility, we used BaMV viral RNA and satBaMV RNA (BSF4) or chimeric satBaMV RNA (BSF4:NbSul and BSF4:NbPDS) as the inocula. The induced mosaic symptoms, silencing phenotypes in WT and *rdr6 N. benthamiana* (Figure S1A), and the analyses by Northern blot (Figure S1B, C and D) or real-time RT-PCR (Figure S1E and F) revealed that *rdr6 N.*

benthamiana plants were hypersusceptible to BaMV and satBaMV in newly emerged leaves and showed enhanced satBaMV-induced gene silencing. In conclusion, the efficiency of satBaMV-induced gene silencing was enhanced in *rdr6 N. benthamiana* regardless of agro-infiltration or mechanical inoculation.

SatBaMV vector suppresses GFP transgene expression in *N. benthamiana*

Because endogenous genes and transgenes may differ in their susceptibility to gene silencing (Fagard and Vaucheret, 2000; Ruiz *et al.*,1998), we tested whether the satBaMV-based vector could suppress the expression of a transgene in *N. benthamiana*. pKF4:GFP was co-infiltrated with pKB into *GFP* transgenic *N. benthamiana* 16c seedlings. Two weeks later, GFP fluorescence was absent in systemic leaves (Figure 4a, 5). It appeared red under ultraviolet (UV) radiation as a result of *GFP* silencing. In contrast to *GFP* transgenic*N. benthamiana*, co-infiltration with the satBaMV vector and pKB (Figure 4a, 4) or pKB alone (Figure 4a, 3) retained the green fluorescence. Moreover, 35S-GFP agro-infiltration in 16c plants could induce efficient *GFP* silencing (Voinnet and Baulcome, 1997). Consequently, 16c plants agro-infiltrated with pKF4:GFP alone (Figure 4a, 6) also showed reduced green fluorescence in systemic leaves, but the silencing effect was less than with pKF4:GFP co-infiltration with pKB (Figure 4a, 5 vs. 6). This silencing phenotype persisted throughout the lifespan of these plants. The mRNA level of cognate *GFP* was decreased by 82.5% and 38.1% in 16c plants co-infiltrated with pKF4:GFP and pKB or pKF4:GFP alone, respectively (Figure 4b).

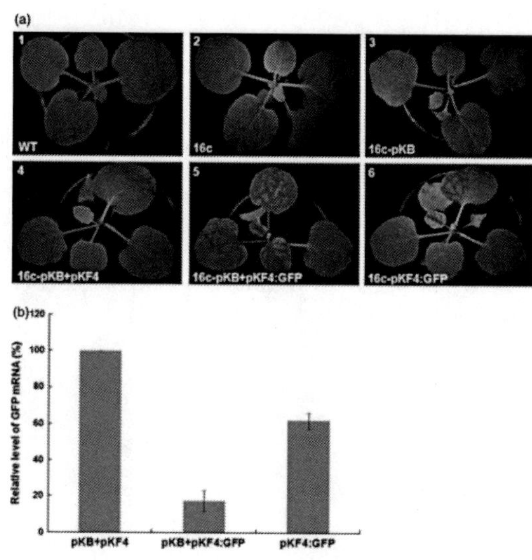

Figure 4. Silencing of transgenes in *Nicotiana benthamiana* with satBaMV-based vector. (a) *GFP*-silenced *N. benthamiana* 16c, nontransgenic*N. benthamiana* (WT) plants appeared red under UV light at 16 dpi (1), whereas nonsilenced control plants showed green fluorescence (2). Phenotypes of *GFP*-silenced *N. benthamiana* 16c plants agro-infiltrated with pKF4:GFP alone (6), pKB alone (3) or co-infiltrated with pKF4 (4), or pKF4:GFP (5) at 16 dpi. (b) Q-RT-PCR analysis of relative mRNA level of *GFP*in nonsilenced and silenced *N. benthamiana* 16c plants. Data are mean ± SD from 3 independent experiments.

Efficient inhibition of endogenous genes in Brachypodium distachyon

To determine whether the satBaMV vector functions to inhibit gene expression in monocots, *B. disatchyon* was used for silencing *Sul* and*PDS*. The 1055- and 441-nt fragments of *Sul* and *PDS* were first amplified by RT-PCR from *B. distachyon* with primers (Supporting information, Table S1) based on *Sul* and *PDS* sequences from complete genome sequences of *B. distachyon* (Vogel *et al.*, 2010). Thus, pCF4:BdSul and pCF4:BdPDS containing a 319-nt fragment of *BdSul* and *BdPDS*, respectively, were generated (Figure 5a). One week after *Chenopodium quinoa* leaves were mechanically co-inoculated with pCB and pCF4, pCF4:BdSul or pCF4:BdPDS, the sap from infected leaves was used as inocula for inducing VIGS in *B. distachyon*. Yellowing and photobleaching developed in newly emerged leaves of the 3-week-old *B. distachyon* line Bd21 after co-inoculation of BaMV and satBaMV:BdSul or satBaMV:BdPDS, silencing *Sul* or*PDS,* respectively (Figure 5b). Plants inoculated with BaMV alone or co-inoculated with satBaMV never showed silencing phenotypes. At 21 dpi, the mRNA levels of *BdSul* and *BdPDS* in leaf tissues were significantly decreased, by 73.2% and 66.5%, in co-inoculated satBaMV:BdSul and satBaMV:BdPDS plants, respectively, as compared with satBaMV co-infected plants (Figure 5c).

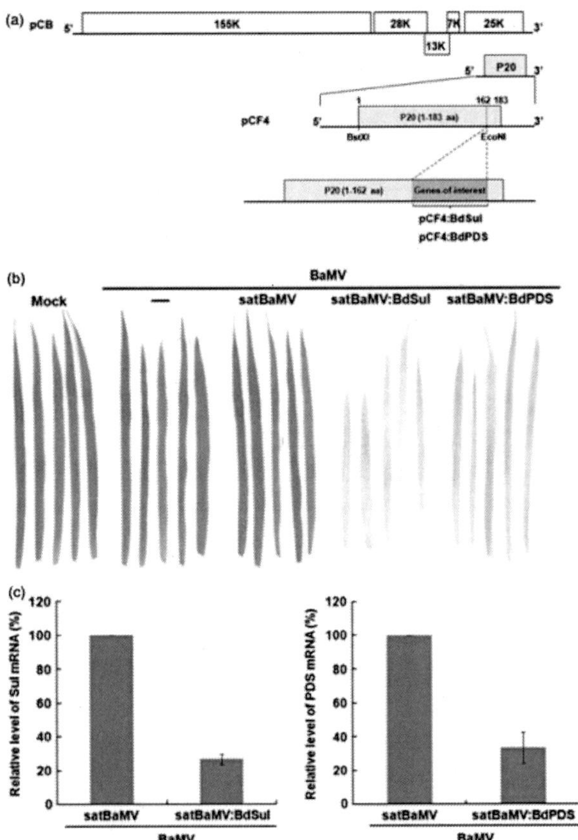

Figure 5. Silencing of *Sul* and *PDS* in *Brachypodium distachyon* with satBaMV-based vector. (a) Genome organization of BaMV, satBaMV expression vectors and satBaMV derivatives with partial *Sul* and *PDS*sequences. (b) Phenotypes from silencing *Sul* and *PDS* with satBaMV-mediated VIGS at 21 dpi. (c) Q-RT-

PCR analysis of relative mRNA levels of*Sul* and *PDS* in WT and silenced *B. distachyon* plants. Vector control value was set to 100%. Data are mean ± SD from 3 independent experiments.

BaMV vector silencing in *N. benthamiana* and *B. distachyon*

Previously, BaMV was developed as a versatile vector for the expression of foreign genes and production of vaccines in plants (Chen *et al.*,2012; Lin *et al.*, 2004; Yang *et al.*, 2007). We further examined its development as a gene-silencing vector. The fragments of*N. benthamiana Sul* and *PDS* were constructed in the BaMV-based vectors (pKBVs) to generate pKBV:NbSul and pKBV:NbPDS, respectively (Figure 6a). *N. benthamiana* plants agro-infiltrated with pKBV:NbSul or pKBV:NbPDS showed yellowing and photobleaching in systemic leaves at 10–14 dpi (Figure 6b). In contrast, plants agro-infiltrated with pKBV vector alone showed only mild mosaic symptoms (Figure 6b). Northern blot analysis confirmed that all viral RNAs, BaMV, BaMV:NbSul and BaMV:PDS accumulated in inoculated and systemic leaves (Figure 6c). The mRNA levels of *Sul* and *PDS* were greatly decreased in silenced leaves (Figure 6d). Thus, a BaMV-based vector could induce VIGS as effectively as a satBaMV-based vector.

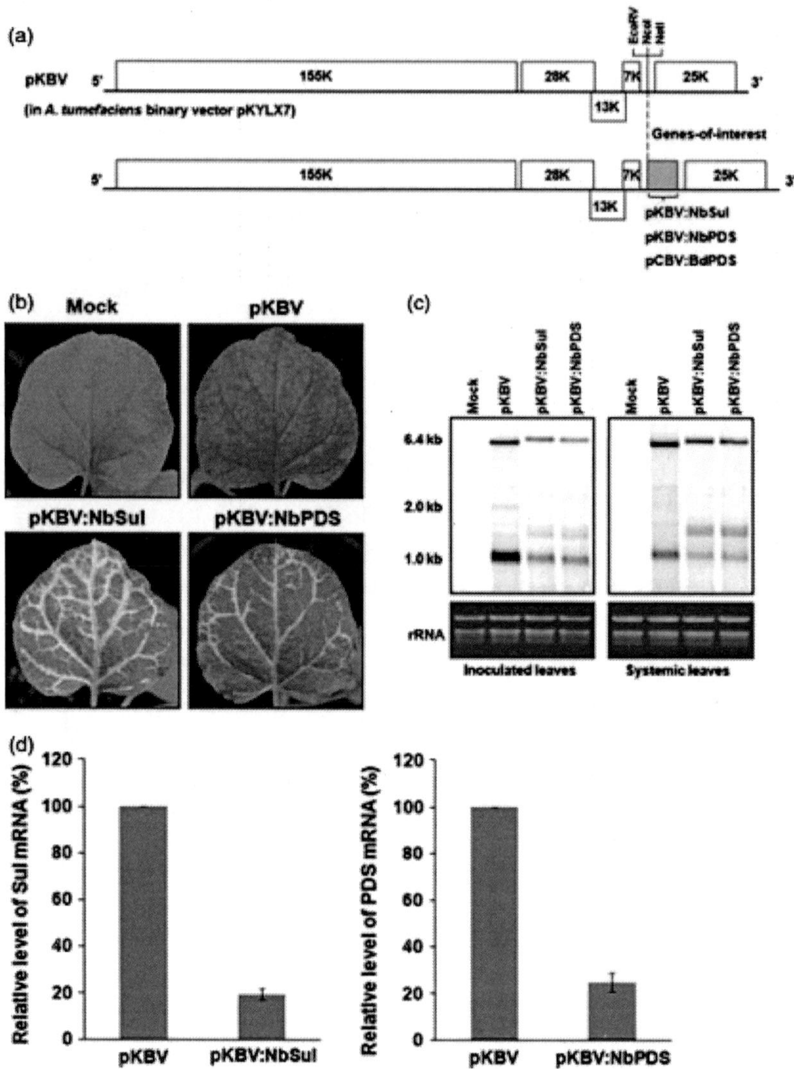

Figure 6. Gene silencing in *Nicotiana benthamiana* with BaMV-based vector carrying *Sul* and *PDS* inserts. (a) Genome organization of BaMV and BaMV-derived constructs to silence *Sul* and *PDS*. The engineered restriction enzyme sites are indicated. (b) Phenotypes of plants agro-infiltrated with a binary vector pKYLX7 only (Mock), pKBV, pKBV:NbSul, pKBV:NbPDS with BaMV-induced gene silencing at 14 dpi. (c) Northern blot analysis of BaMV accumulation in inoculated and systemic leaves of *N. benthamiana* agro-infiltrated with pKBV, pKBV:NbSul, and pKBV:NbPDS at 16 dpi. Blots were hybridized

with DIG-labelled RNA complementary to the 3′ UTR of BaMV RNA (Lin *et al.*, 1993) as described in Figure 1c. (d) Q-RT-PCR analysis of relative mRNA levels of *Sul* and *PDS* in WT and silenced *N. benthamiana* plants. Vector control value was set to 100%. Data are mean ± SD from 3 independent experiments.

To test whether the BaMV vector could silence gene expression in monocots, we constructed pCBV:BdPDS carrying a *PDS* gene fragment from *B. distachyon* (Figure 6a). Again, mechanically inoculated *C. quinoa* leaves were used as inocula for inducing VIGS in *B. distachyon*. The infected *B. distachyon* plants showed streaking and photobleaching in newly emerged leaves at 40 dpi (Figure 7a). BaMV and BaMV:BdPDS levels were induced in inoculated and systemic leaves of *B. distachyon* at 16 dpi (Figure 7b). Further, the mRNA levels of*PDS* were reduced with pCBV:BdPDS knockdown to about 61.9% of that of the control (Figure 7c).

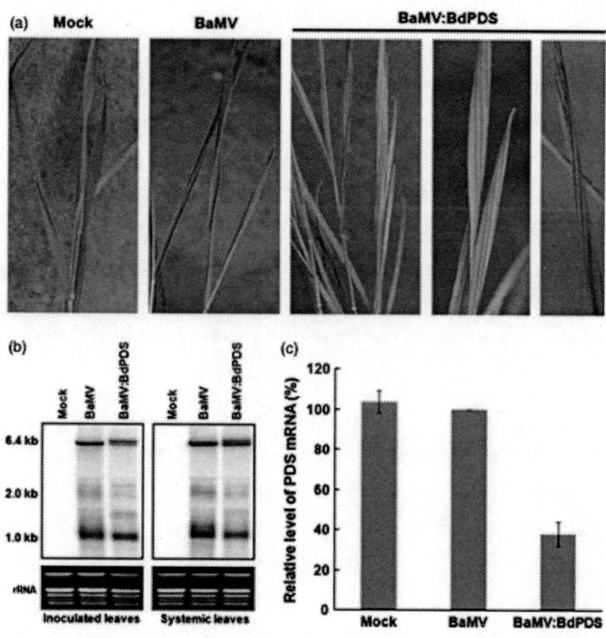

Figure 7. BaMV-based vector efficiently silenced *PDS* gene in *Brachypodium distachyon*. (a) Phenotypes resulting from silencing *PDS* gene by BaMV-mediated VIGS at 40 dpi. (b) Northern blot analyses of BaMV accumulation in infected and systemic leaves of *B. distachyon* inoculated with BaMV or BaMV:BdPDS at 16 dpi. (c) Q-RT-PCR analysis of relative mRNA level of *PDS*in systemic leaves

of WT and silenced *B. distachyon*. Vector control value was set to 100%. Data are mean ± SD from 3 independent experiments.

Compromised BaMV-induced gene silencing in *rdr6 N. benthamiana*

As described above, satBaMV-induced gene silencing was enhanced in *rdr6 N. benthamiana*. We next tested whether this enhancement also occurred with the BaMV-VIGS vector. WT and *rdr6* plants were agro-infiltrated with pKBV, pKBV:NbSul or pKBV:NbPDS. At 16 dpi, WT and *rdr6* plants infiltrated with pKBV:NbSul and pKBV:NbPDS showed a yellow-white phenotype as compared with mock inoculation (Figure S2A, 3–4 vs. 1; 7–8 vs. 5). In contrast to WT plants, the systemic and newly emerged leaves of *rdr6* with pKBV:NbSul and pKBV:NbPDS infection were largely free of the *Sul* and *PDS* silencing phenotype (Figure S2A, 7–8 vs. 3–4). Northern blot analysis with BaMV-specific probes confirmed that the RNA levels of BaMV and chimeric BaMV were slightly higher in inoculated leaves of *rdr6* than in WT plants but nearly equal in the 6th systemic leaves (Figure S2B-C, lanes 7–8 vs. lanes 3–4). As in the satBaMV-VIGS system shown in Figure 3, the expression of pKBV, pKBV:NbSul and pKBV:NbPDS accumulated substantially, with detectable RNA and siRNAs level of BaMV in 8th systemic leaves of *rdr6* plants but barely detectable levels in the WT (Figure S2D, lanes 6–8 vs. lanes 2–4). Likewise, the accumulation of BaMV siRNA was lower in inoculated and 6th systemic leaves of *rdr6* plants than in WT leaves (Figure S2B-C, lanes 2–4 vs. lanes 6–8). These silencing phenotypes well matched a lower mRNA accumulation of *Sul* and *PDS* in WT than *rdr6* leaves (Figure S2A, E and F). The silencing efficiency of pKBV:NbSul and pKBV:NbPDS was great in WT plants, at 73.9% and 67.4%, respectively, than in*rdr6* plants, at 41.2% and 38.6%, respectively (Figure S2E and F). Therefore, the efficiency of BaMV-induced gene silencing was compromised in *rdr6 N. benthamiana*.

Silencing of 2 genes concurrently with BaMV and satBaMV-VIGS vectors

As shown previously, BaMV- and satBaMV-based vectors could effectively silence endogenous genes in *N. benthamiana* and*B.*

distachyon. To examine the silencing of 2 genes simultaneously, we co-inoculated *GFP* transgenic *N. benthamiana* 16c plants with BaMV and satBaMV vectors carrying *Sul* and *GFP*, respectively (Figure 8a). At 16 dpi, the *GFP* transgenic 16c plants co-infiltrated with pKBV:NbSul and pKF4:GFP showed the loss of green fluorescence along the vein under UV radiation (Figure 8a, 1d) and the photobleaching phenotype of *Sul* silencing appeared near the vein under normal light (Figure 8a, 2d). However, 16c plants infiltrated with pKBV:NbSul alone (Figure 8a, 1b) or co-infiltrated with pKF4 (Figure 8a, 1c) retained the green fluorescence under UV radiation and showed the *Sul* silencing phenotype (Figure 8a, 2b and 2c). Similarly, 16c plants co-infiltrated with pKBV and pKF4:GFP lost green fluorescence (Figure 8a, 1h) but showed no *Sul* silencing phenotype (Figure 8a, 2h). In contrast, plants co-infiltrated with pKBV and pKF4 (Figure 8a, 1g) or pKBV alone (Figure 8a, 1f) retained the green fluorescence, with no *Sul* silencing phenotype (Figure 8a, 2f and 2g). Northern blot analysis confirmed the accumulation of BaMV, satBaMV and their derived constructs in inoculated and systemic leaves (Figure 8b). The mRNA levels of *GFP* and *Sul* were lower, by 87.7% and 77.9%, respectively, in 16c plants co-infiltrated with pKBV:NbSul and pKF4:GFP than in control plants co-infiltrated with pKF4 and pKBV vector for *GFP* gene normalization or with pKBV vector alone for *Sul*gene normalization at 21 dpi (Figure 8c). However, 16c plants infiltrated with pKBV:NbSul or co-infiltrated with pKBV:NbSul and pKF4 showed no *GFP* reduction but lower *Sul* mRNA levels, by 75.1% and 73.4%, respectively, than in control plants co-infiltrated with pKF4 and pKBV vector (Figure 8c). Nevertheless, control plants co-infiltrated with pKBV empty vector and pKF4:GFP showed decreased mRNA levels of *GFP*, by 84.3% (Figure 8c). Thus, *Sul* and *GFP* genes could be concurrently effectively silenced with BaMV and satBaMV vector systems.

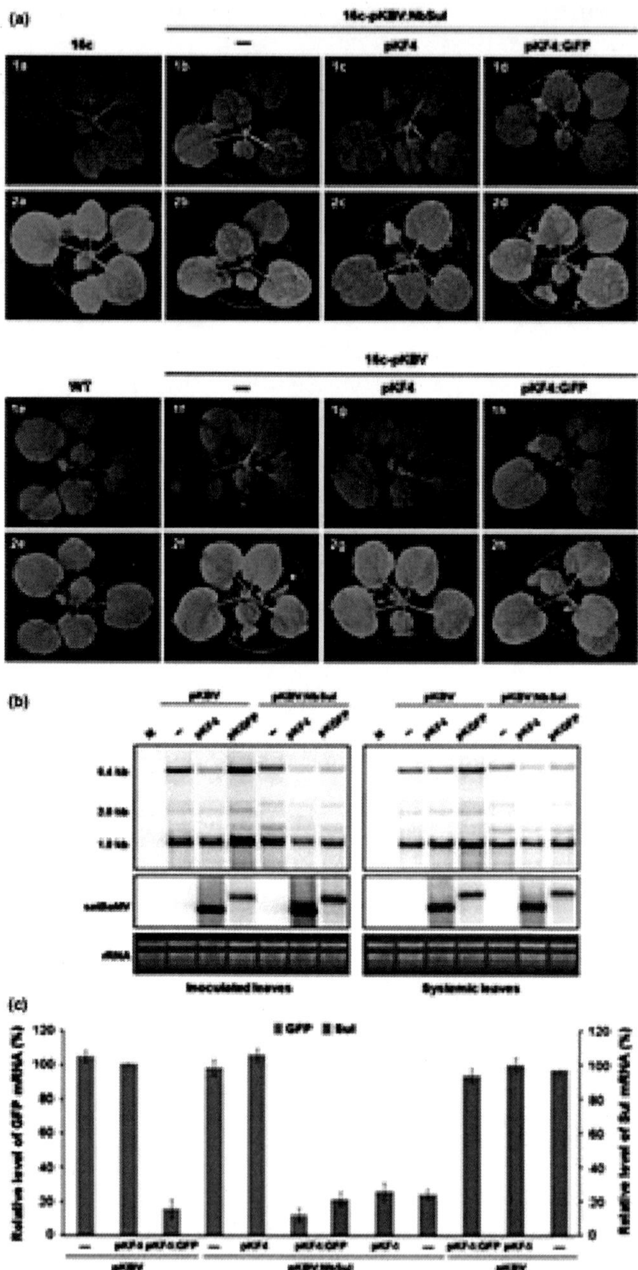

Figure 8. *Sul* and *GFP* were concurrently silenced with BaMV and satBaMV dual gene silencing. (a) Phenotypes of *GFP* transgenic *Nicotiana benthamiana*

16c plants agro-infiltrated with mock, pKBV, pKBV:NbSul, or co-infiltrated pKBV with pKF4 or pKF4:GFP, or pKBV:NbSul with pKF4 or pKF4:GFP at 16 dpi. (b) Northern blot analyses of BaMV and satBaMV levels in infected and systemic leaves of *N. benthamiana* 16c plants. (c) Q-RT-PCR analysis of relative mNRA levels of *GFP* and *Sul* in *GFP* transgenic*N. benthamiana* 16c plants after agro-infiltration with the indicated plasmids. Vector (pKBV for *Sul* and pKBV+pKF4 for *GFP*) control value was set to 100%. Data are mean ± SD from 3 independent experiments.

Stability of BaMV- and satBaMV-based VIGS vectors

The reduction or recovery of silencing phenotypes was frequently observed in many VIGS systems, mainly because the VIGS vectors became targets of RNA degradation that led to the instability of the vectors (Voinnet, 2001). To test the stability of BaMV- and satBaMV-based VIGS vectors, we performed RT-PCR and sequence analysis on the chimeric virus or satRNA progenies recovered from systemic leaves, including the recovered and un-silenced regions. As shown in Figure S3, the majority (>85% and >65% in *N. benthamiana* and*B. distachyon*, respectively) of the progenies of chimeric BaMV or satBaMV retained the original inserts at 16 dpi, as indicated by the slower electrophoresis mobility of the PCR products. Sequence analyses of the slower moving PCR products confirmed the sequence integrity of the progenies of the chimeric constructs and revealed that the minor, faster-moving fragments represented the respective vectors with the inserts deleted (Figure S3A, a and d; S3B, a and c). The stabilities of the chimeric constructs were affected by the inserts (Sul or PDS), the type of vectors (BaMV or satBaMV) and the host plants (*N. benthamiana* or *B. distachyon*). For insert types, *Sul* gene fragments in BaMV- or satBaMV-based vectors were more stable in *N. benthamiana* leaves (Figure S3A, a3 and d8). In contrast, the *PDS* gene fragments were less stable both in *N. benthamiana* (Figure S3A, a4 and d9) and *B. distachyon* (Figure S3B, a3 and c8) in both vectors. For vector types,*Sul* gene fragments were more stable in satBaMV-based constructs. Concerning host plants, both BaMV- and satBaMV-based vectors were more stable in *N. benthamiana* than in *B. distachyon* (Figure S3A vs. S3B). In general, the results indicated that the BaMV- and satBaMV-based VIGS vectors maintained appreciable stabilities in both monocot and dicot host plants with different target gene fragments.

DISCUSSION

In this study, we first demonstrated that a satBaMV-based vector can function in VIGS in monocot and dicot plants. This vector is particularly useful for dealing with genes such as *Hsp70* and *Hsp90* that cause adverse vital effects in plants. In addition, BaMV and satBaMV were hyperaccumulated and the efficiency of satBaMV-induced gene silencing was enhanced in *rdr6 N. benthamiana*. With a BaMV-induced gene-silencing system, a dual gene-silencing vector based on BaMV and satBaMV could silence 2 genes simultaneously, and is applicable with both monocot and dicot plants for functional genomics studies.

SatBaMV-VIGS Vector in monocot and dicot plants

Previously, two-component gene-silencing systems such as satellite virus (Gosselé *et al.*, 2002), DI RNA (Hou and Qiu, 2003) and satellite DNA (Huang *et al.*, 2009; Tao and Zhou, 2004) have been modified as VIGS vectors. However, the use of satRNA as a gene-silencing vector has not yet been established, although satBaMV has been developed as a foreign protein expression vector in plants (Lin*et al.*, 1996). In this study, we developed satBaMV and BaMV vectors in both *N. benthamiana* and *B. distachyon*. With the exception of the BMV (Pacak *et al.*, 2010b) and BSMV (Yuan *et al.*, 2011) VIGS system, few VIGS vectors can be used in monocot and dicot plants.

Furthermore, the satBaMV and BaMV-VIGS systems can induce gene silencing by 'one-step' mechanical inoculation of plasmid DNA or agro-infiltration. In contrast, the BSMV VIGS system requires the laborious process of synthesis of viral RNA by *in vitro* transcription. Similar 'one-step' vector systems involving mechanical inoculation of plasmid DNA were demonstrated in the VIGS vectors TYMV (Pflieger*et al.*, 2008) and BPMV (Zhang *et al.*, 2009) in the dicots *Arabidopsis* and soybean, respectively. Because our plasmid DNA only needs to be subcloned into binary vectors for *Agrobacterium*-mediated transient expression in plant cells, it can be easily applied at low cost. In addition, it is an efficient way to inoculate a large number of plants and has potential for high-throughput plant functional genomics. Furthermore, a satBaMV-induced gene silencing system is favourable for silencing essential endogenous genes required for plant growth and development. As shown in Figure 2e and f, TRV silencing of development-related genes such as *Hsp90* and *Hsp70* caused adverse vital effects in plants, but satBaMV-silenced plants continued to grow. By comparison to the TRV-VIGS vector, the satBaMV vector induces less severe developmental abnormalities, possibly because of the nature of the satRNA, which normally competes for replicase complexes with its helper virus (Hu *et al.*, 2009). Thus, the VIGS systems described in this study may have wider applications.

Our data showed that pCF4:Sul-N3 and pCF4:Sul-N4 triggered statistically more efficient silencing effect than did other satBaMV constructs, as evaluated by *Sul* mRNA accumulation (Figure 1d). One possibility was that, in pCF4:Sul-N3 and pCF4:Sul-N4, the N-terminal structural integrity of the P20 protein was retained, which would induce better gene-silencing efficiency. According to computer prediction, P20 structurally resembles the jelly roll structure of the CP encoded by *Satellite panicum mosaic virus,* which is a typical eight-stranded β-sheet conformation (Ban and McPherson, 1995; Liu *et al.*, 1997). Seven and eight β-sheets were retained in pCF4:Sul-N3 and pCF4:Sul-N4, respectively, and might have contributed to the higher silencing efficiency. However, the accumulation of chimeric satBaMV RNA was not detected in *N. benthamiana* leaves co-inoculated with pCF4:Sul-N5 and pCB, which retained the whole P20 sequence (Figure 1c). The lack of apparent viability of pCF4:Sul-N5 could be resulted from the insertion of *Sul* gene fragment in between the P20 ORF and the 3′ UTR of satBaMV RNA, which would interfere with the formation of a 3′ terminal secondary structure, as predicted by Mfold program (Zuker, 2003) and confirmed by enzymatic structural probing (Hsu, Y. H., unpublished data). On the other hand, the difference in*Sul* mRNA accumulation was not statistically significant either in the presence (pCF4:Sul-N1 and pCF4:Sul-N2) or the absence (pCF4:Sul-C1 and pCF4:Sul-C2) of the N-terminal region of P20, suggesting that the N-terminal 1–69 amino acids of p20 did not affect the silencing efficiency. Likewise, previous data showed that P20 of satBaMV was not essential for satBaMV replication and could be replaced with a foreign gene (Lin *et al.*, 1996). Thus, whether the N-terminal region of P20 enhanced the silencing efficiency remained to be elucidated.

Enhanced satBaMV-induced gene silencing but compromised BaMV-driven VIGS in rdr6 N. benthamiana

In general, primary siRNAs derived from DCL-mediated cleavage of viral dsRNA intermediates are the initiators of RNA silencing. The amplification of secondary siRNAs by plant-encoded RDR6 is required for maintaining RNA silencing. In this study, NbRDR6-silenced plants were hypersusceptible to BaMV and satBaMV in newly emerged leaves (Figure 3a, 7–8). Similar results were demonstrated for HSVd (Gómez *et al.*, 2008), PSTVd (Di Serio *et al.*, 2010), PVX (Qu *et al.*, 2005; Schwach *et al.*, 2005), PVY, PPV and CMV with the Y satellite but not in TRV, TMV, *Turnip crinkle virus* (TCV) and CMV alone (Schwach *et al.*, 2005). Thus, invasion of BaMV and satBaMV into newly emerging leaves of infected plants may have been restricted by NbRDR6. Interestingly, although *rdr6 N. benthamiana* can hyperaccumulate PVX and PPV, the efficiency of PVX- and PPV-driven VIGS was compromised (Vaistij and Jones, 2009), similar to BaMV-driven VIGS (Figure S2). However, efficiency of satBaMV-induced gene silencing was enhanced in *rdr6* plants (Figure 3). Thus, the gene-silencing mechanism may differ between RNA viruses and satRNAs.

Indeed, the RNA levels of BaMV and satBaMV were slightly higher in inoculated and 6th systemic leaves of *rdr6* than WT plants, but their siRNAs accumulated dissimilarly (Figure 3b–c). The siRNA level of BaMV was lower in inoculated and 6th systemic leaves of *rdr6* plants than in WT plants. In contrast, the satBaMV siRNA level was higher in 6th systemic leaves (Figure 3c). The secondary but not primary siRNAs of BaMV and satBaMV in systemic leaves might be regulated by RDR6 differently. As well, the newly emerged leaves of WT and*rdr6* plants showed strikingly distinct accumulation of viral RNAs and siRNAs (Figure 3d). The results verified the crucial role of RDR6 in limiting virus invasion into emerged leaves (Schwach *et al.*, 2005). Moreover, satRNA persisted to generate dsRNA intermediates by replicase complexes, then cleaved into primary siRNA by DCL proteins and itself may not be targeted by RDR6 for producing secondary siRNAs. For RNA viruses such as PVX and PPV, RDR6 is required to produce secondary siRNAs that trigger more efficient RNA silencing (Vaistij and Jones, 2009). Our Figure S2B–C show lower siRNA level of BaMV in inoculated and 6th systemic leaves of *rdr6* than in WT plants. Because BaMV can hyperaccumulate in the newly emerged leaves of *rdr6* plants, the siRNA level of BaMV would be higher than in WT plants (Figure S2D). Consequently, BaMV is RDR6-dependent for secondary siRNA production, thus resulting in decreased VIGS in*rdr6* plants. However, a higher titre of the helper virus accumulating in *rdr6* plants would enhance the satRNA replication, thus leading to increased VIGS because of increased RDR6-independent production of sat-siRNAs. Thus, RDR6 may affect BaMV and satBaMV invasion into newly emerged leaves and also mediate the production of virus-derived secondary siRNAs but not siRNAs from satRNAs.

Dual gene-silencing vector system

We demonstrated that BaMV and satBaMV-VIGS vectors could silence 2 genes simultaneously. Such a dual gene-silencing vector has several advantages, they are follows: 1) Double mutant plants could be easily obtained without treatment with chemical mutagenesis, T-DNA insertion or transposon insertion. 2) The vectors can be used to identify the interaction between genes and pathways such as metabolism or disease resistance. 3) A silencing phenotype can serve as a marker for fast confirmation of silencing efficiency. For instance, the BaMV helper vector silenced *Sul*, with an observable phenotype, and the satBaMV-VIGS vector silenced *GFP* simultaneously (Figure 8a). To our knowledge, this is the first demonstration of 2 genes silenced concurrently by a virus and satRNA VIGS system. With the VIGS vectors such as BPMV (Zhang *et al.*, 2009), BSMV (Cakir and Tör, 2010), CbLCV (Turnage *et al.*, 2002), TbSCV DNA1 (Huang*et al.*, 2009) and TYCCNV DNAβ (Tao and Zhou, 2004), simultaneous silencing of 2 genes relied on the same viral vector. Our dual transient gene-silencing system could be used for both monocot and dicot plants for high-throughput functional genomics studies.

Experimental Procedures

Plant materials and growing conditions

Seeds of *GFP* transgenic *N. benthamiana* 16c line harbouring a single copy of the modified GFP (mGFP) transgene as described (Ruiz*et al.*, 1998), and seeds of *rdr6* transgenic *N. benthamiana* with *RDR6* silenced by RNA interference with a hairpin construct (Schwach*et al.*, 2005) were obtained from Dr. D. C. Baulcombe (University of Cambridge, UK). *N. benthamiana* and the *B. distachyon* Bd21 line (kindly provided by Dr. A. O. Jackson, University of California, Berkeley, CA, USA) were used for VIGS of endogenous genes with BaMV and satBaMV vectors. *N. benthamiana* plants were grown at 28 °C with a 16-h light/8-h dark photoperiod in a greenhouse. *B. distachyon* plants were grown in a growth chamber at 25 °C under a 16-h light/8-h dark cycle. *C. quinoa* plants were grown in a greenhouse under natural conditions.

Construction of satBaMV, BaMV and TRV vectors

Plasmids pCB and pCF4 are infectious clones of BaMV-S and satBaMV, respectively, driven by an 35S promoter (Lin *et al.*, 2004; Vijayapalani *et al.*, 2012). To construct the gene-silencing vector, a vector pCBV was derived from pCB, and a duplicated subgenomic promoter and multiple cloning sites (*Eco*RV-*Nco*I-*Not*I) were inserted between the 2 ORFs of 7 and 25 kDa in the BaMV genome (Figure 6a). For agro-infiltration, pCB and pCBV were subcloned into the *Agrobacterium tumefaciens* binary vector pKYLX7 (Schardl *et al.*,1987) with the use of *Hin*dIII and *Sac*I and named pKB and pKBV, respectively. Plasmid pCF4 was subcloned into the pKn vector (the rbsc terminator of pKYLX7 vector replaced with a nos terminator) with *Hin*dIII and *Bgl*II and referred to as pKF4.

Details for the construction of satBaMV, BaMV and TRV vectors are described in the Supporting information (Method S1). All plasmid DNA construction followed a standard protocol (Sambrook and Russell, 2001), and all clones were confirmed by sequencing. The primers for satBaMV, BaMV and TRV-VIGS construction are in the Supporting information (Tables S1 and S2).

Plant inoculation

For mechanical inoculation, the constructed vectors were purified from large-scale cultures of *Escherichia coli* DH10B with the use of a NucleoBond® plasmid Maxi kit and then mechanically inoculated into *C. quinoa* and *N. benthamiana*. For *B. distachyon*, leaves of*C. quinoa* with symptoms at 7 dpi were homogenized in 5 volumes of 0.01 M sodium phosphate buffer (pH 7.0) as inoculum and reinoculated into *C. quinoa* as a positive control. The infected leaves were used as inocula for inducing VIGS in *B. distachyon*.

For VIGS assay by agro-infiltration, pKB and pKF4 or their derivatives were introduced into *A. tumefaciens* strain GV3850 by electroporation at 25 µF, 400 Ω, for a 2.5-KV pulse. A 5-mL culture was grown overnight at 28 °C in the appropriate antibiotic selection medium (50 mg/L ampicillin, 50 mg/L tetracycline and 10 mg/L kanamycin). The *Agrobacterium* cells were harvested after overnight culture and resuspended in infiltration media (10 mM $MgCl_2$, 10 mM MES, 200 µM acetosyringone), adjusted to an optical density at 600 nm of 1.0 and left at room temperature for 2–3 h. For the satBaMV-VIGS system, separate cultures containing pKB and pKF4 or pKF4-derived constructs were mixed at a 1:1 ratio. The culture was then infiltrated into the underside of leaves of *N. benthamiana* (2- to 3-leaf stage) by the use of a 1-mL syringe without a needle. The infiltrated plants were maintained in the greenhouse at 28 °C and 16-h light/8-h dark to induce VIGS.

RNA extraction and Northern blot analyses

Total RNA was extracted from plants as described (Verwoerd *et al.*, 1989). The RNA levels of BaMV, satBaMV and their chimera in inoculated plants were assayed by Northern blot hybridization (Lin *et al.*, 1996). BaMV and satBaMV RNAs were probed with digoxigenin (DIG)-labelled *in vitro* transcripts from cloned *Hin*dIII-linearized pBaHB (Lin *et al.*, 1993) or *Eco*NI-linearized pBSHE (Lin *et al.*, 1996; Vijayapalani *et al.*, 2012) and RNA probes specific for BaMV 3'UTR and satBaMV, respectively. The hybridized membranes were immuno-detected with anti-DIG Fab fragments coupled with alkaline phosphatase and visualized with a chemiluminescent substrate, CSPD. The banding signals were scanned and quantified by the use of PhosphorImager (Fujifilm LAS-4000, Multi Gauge version 3.1, Fujifilm). The low-molecular-weight RNA (5 µg) for each sample was separated on 8 M urea-15% polyacrylamide gel and then transferred to a nylon membrane via capillary blotting by standard methods. Blotted RNA was hybridized overnight at 40 °C with a

^{32}P-labelled BaMV and satBaMV RNA probe in hybridization buffer (50% formamide, 6X SSC, 5X Dehard's reagent, 100 μg denature-fragmented salmon sperm DNA and 0.5% SDS). The membrane was washed twice with 2X SSC (150 mM NaCl and 15 mM sodium citrate) and 0.1% SDS for 20 min at 40 °C. The banding signals were scanned and quantified by the use of PhosphorImager (Fujifilm LAS-2500, Multi Gauge version 3.1, Fujifilm).

Quantitative real-time RT-PCR (Q-RT-PCR)

Q-RT-PCR involved use of a Rotor-Gene 3000. The total RNA concentration was measured by the use of micro-spectrophotometry. Total RNA used in first-strand cDNA synthesis was treated with RNase-free DNase I to avoid DNA contamination. The first-strand cDNA was generated by reverse transcription using 1 μg of total RNA as the template and oligo $d(T)_{18}$ primer. For Q-RT-PCR, primer sequences are listed in Supporting information (Table S3). The level of the actin gene of *N. benthamiana* and *B. distachyon* was determined for the normalization of specifically silenced genes.

Photography and GFP imaging

Phenotypes of endogenous gene silencing were assessed visually, and plants were photographed with the use of a Nikon digital camera, D70. GFP fluorescence in whole plants was observed with the use of a hand-held 100 W long-wave UV lamp, and photographs were taken by the use of a Nikon D70 digital camera fitted with UV and yellow-green filters. Image processing involved the use of Adobe Photoshop CS4.

ACKNOWLEDGEMENTS

We are grateful to Dr. David C. Baulcombe (University of Cambridge, UK) for the TRV-VIGS vector and seeds of *GFP* transgenic*N. benthamiana* line 16c and *rdr6*. We also thank Professor Andrew O. Jackson (University of California, Berkeley) for *B. distachyon* seeds and Dr. Ching-Hsiu Tsai (National Chung Hsing University, Taiwan) for pTRV2/GFP. This research was supported by the National Science Council, Taiwan (Grants NSC-97-2752-B-005-001-PAE) and Aim for the Top University Project of the National Chung Hsing University by the Ministry of Education, Taiwan (CC00313/100S0501).

REFERENCES

1. Ban, N. and McPherson, A. (1995) The structure of satellite panicum mosaic virus at 1.9 Å resolution. *Natl. Struct. Biol.* **2**, 882–890.
2. Baulcombe, D.C. (1999) Gene silencing: RNA makes RNA makes no protein. *Curr. Biol.* **9**, R599–R601.
3. Burch-Smith, T.M., Anderson, J.C., Martin, G.B. and Dinesh-Kumar, S.P. (2004) Applications and advantages of virus-induced gene silencing for gene function studies in plants. *Plant J.* **39**, 734–746.
4. Cakir, C. and Tör, M. (2010) Factors influencing barley stripe mosaic virus-mediated gene silencing in wheat. *Physiol. Mol. Plant Pathol.* **74**, 246–253.
5. Chen, T.H., Chen, T.H., Hu, C.C., Liao, J.T., Lee, C.W., Liao, J.W., Lin, M.Y., Liu, H.J., Wang, M.Y., Lin, N.S. and Hsu, Y.H. (2012)Induction of protective immunity in chickens immunized with plant-made chimeric bamboo mosaic virus particles expressing very virulent Infectious bursal disease virus antigen. *Virus Res.* **166**, 109–115.
6. Dalmay, T., Hamilton, A.J., Rudd, S., Angell, S. and Baulcombe, D.C. (2000) An RNA-dependent RNA polymerase gene in*Arabidopsis* is required for posttranscriptional gene silencing mediated by a transgene but not by a virus. *Cell* **101**, 543–553.
7. Di Serio, F., Martinez de Alba, A.E., Navarro, B., Gisel, A. and Flores, R. (2010) RNA-dependent RNA polymerase 6 delays accumulation and precludes meristem invasion of a viroid that replicates in the nucleus. *J. Virol.* **84**, 2477–2489.
8. Dong, Y., Burch-Smith, T.M., Liu, Y., Mamillapalli, P. and Dinesh-Kumar, S.P. (2007) A ligation-independent cloning tobacco rattle virus vector for high-throughput virus-induced gene silencing identifies roles for NbMADS4-1 and -2 in floral development. *Plant Physiol.* **145**, 1161–1170.
9. Dunoyer, P., Himber, C. and Voinnet, O. (2006) Induction, suppression and requirement of RNA silencing pathways in virulent*Agrobacterium tumefaciens* infections. *Nat. Genet.* **38**, 258–263.
10. Fagard, M. and Vaucheret, H. (2000) (Trans) gene silencing in plants: how many mechanisms? *Annu Rev. Plant Physiol. Plant Mol. Biol.* **51**, 167–194.
11. Garcia-Ruiz, H., Takeda, A., Chapman, E.J., Sullivan, C.M., Fahlgren,

N., Brempelis, K.J. and Carrington, J.C. (2010) Arabidopsis RNA-dependent RNA polymerases and dicer-like proteins in antiviral defense and small interfering RNA biogenesis during turnip mosaic virus infection. *Plant Cell* **22**, 481–496.

12. Godge, M.R., Purkayastha, A., Dasgupta, I. and Kummar, P.P. (2008) Virus-induced gene silencing for functional analysis of selected genes. *Plant Cell Rep.* **27**, 209–219.
13. Gómez, G., Martínez, G. and Pallás, V. (2008) Viroid-induced symptoms in *Nicotiana benthamiana* plants are dependent on RDR6 activity. *Plant Physiol.* **148**, 414–423.
14. Gosselé, V., Faché, I., Meulewaeter, F., Cornelissen, M. and Metzlaff, M. (2002) SVISS - a novel transient gene silencing system for gene function discovery and validation in tobacco plants. *Plant J.* **32**, 859–866.
15. Hemmer, O., Oncino, C. and Fritsch, C. (1993) Efficient replication of the in vitro transcripts from cloned cDNA of tomato black ring virus satellite RNA requires the 48K satellite RNA-encoded protein. *Virology* **194**, 800–806.
16. Holzberg, S., Brosio, P., Gross, C. and Pogue, G.P. (2002) Barley stripe mosaic virus-induced gene silencing in a monocot plant.*Plant J.* **30**, 315–327.
17. Hou, H. and Qiu, W. (2003) A novel co-delivery system consisting of a tomato bushy stunt virus and a defective interfering RNA for studying gene silencing. *J. Virol. Methods* **111**, 37–42.
18. Hu, C.C., Hsu, Y.H. and Lin, N.S. (2009) Satellite RNAs and satellite viruses of plants. *Viruses* **1**, 1325–1350.
19. Huang, C., Xie, Y. and Zhou, X. (2009) Efficient virus-induced gene silencing in plants using a modified geminivirus DNA1 component. *Plant Biotechnol. J.* **7**, 254–265.
20. Huang, Y.W., Hu, C.C., Lin, N.S. and Hsu, Y.H. (2010) Mimicry of molecular pretenders: the terminal structures of satellites associated with plant RNA viruses. *RNA Biol.* **7**, 162–171.
21. Koncz, C., Mayerhofer, R., Koncz-Kalman, Z., Nawrath, C., Reiss, B., Redei, G.P. and Schell, J. (1990) Isolation of a gene encoding a novel chloroplast protein by T-DNA tagging in *Arabidopsis thaliana. EMBO J.* **9**, 1337–1346.
22. Kumagai, M.H., Donson, J., della-Cioppa, G., Harvey, D., Hanley, K. and Grill, L.K. (1995) Cytoplasmic inhibition of carotenoid biosynthesis with virus-derived RNA. *Proc. Natl Acad. Sci. USA*, **92**, 1679–1683.
23. Lee, Y.S., Hsu, Y.H. and Lin, N.S. (2000) Generation of subgenomic

RNA directed by a satellite RNA associated with bamboo mosaic potexvirus: analyses of potexvirus subgenomic RNA promoter. *J. Virol.* **74**, 10341–10348.

24. Lin, N.S. and Hsu, Y.H. (1994) A satellite RNA associated with bamboo mosaic potexvirus. *Virology* **202**, 707–714.
25. Lin, N.S., Chai, Y.J., Huang, T.Y., Chang, T.Y. and Hsu, Y.H. (1993) Incidence of bamboo mosaic potexvirus in Taiwan. *Plant Dis.* **77**,448–450.
26. Lin, N.S., Lin, B.Y., Lo, N.W., Hu, C.C., Chow, T.Y. and Hsu, Y.H. (1994) Nucleotide sequence of the genomic RNA of bamboo mosaic potexvirus. *J. Gen. Virol.* **75**, 2513–2518.
27. Lin, N.S., Lee, Y.S., Lin, B.Y., Lee, C.W. and Hsu, Y.H. (1996) The open reading frame of bamboo mosaic potexvirus satellite RNA is not essential for its replication and can be replaced with a bacterial gene. *Proc. Natl Acad. Sci. USA* **93**, 3138–3142.
28. Lin, M.K., Chang, B.Y., Liao, J.T., Lin, N.S. and Hsu, Y.H. (2004) Arg-16 and Arg-21 in the N-terminal region of the triple-gene-block protein 1 of bamboo mosaic virus are essential for virus movement. *J. Gen. Virol.* **85**, 251–259.
29. Liu, Y.Y. and Cooper, J.I. (1993) The multiplication in plants of arabis mosaic virus satellite RNA requires the encoded protein. *J. Gen. Virol.* **74**, 1471–1474.
30. Liu, J.S., Hsu, Y.H., Huang, T.Y. and Lin, N.S. (1997) Molecular evolution and phylogeny of satellite RNA associated with bamboo mosaic potexvirus. *J. Mol. Evol.* **44**, 207–213.
31. Liu, Y., Schiff, M., Marathe, R. and Dinesh-Kumar, S.P. (2002) Tobacco *Rar1, EDS1* and *NPR1/NIM1* like genes are required for *N*-mediated resistance to tobacco mosaic virus. *Plant J.* **30**, 415–429.
32. Liu, Y., Burch-Smith, T., Schiff, M., Feng, S. and Dinesh-Kumar, S.P. (2004) Molecular chaperone Hsp90 associates with resistance protein N and its signaling proteins SGT1 and Rar1 to modulate an innate immune response in plant. *J. Biol. Chem.* **279**,2101–2108.
33. Lu, R., Malcuit, I., Moffett, P., Ruiz, M.T., Peart, J., Wu, A.J., Rathjen, J.P., Bendahmane, A., Day, L. and Baulcombe, D.C. (2003) High throughput virus-induced gene silencing implicates heat shock protein 90 in plant disease resistance. *EMBO J.* **22**, 5690–5699.
34. Pacak, A., Geisler, K., Jorgensen, B., Barciszewska-Pacak, M., Nilsson, L., Nielsen, T.H., Johansen, E., Gronlund, M., Jacobsen, I. andAlbrechtsen, M. (2010a) Investigations of barley stripe mosaic virus as a gene silencing vector in barley roots and in*Brachypodium*

distachyon and oat. *Plant Methods* **6**, 26.

35. Pacak, A., Strozycki, P.M., Barciszewska-Pacak, M., Alejska, M., Lacomme, C., Jarmolowski, A., Szweykowska-Kulinska, Z. andFiglerowicz, M. (2010b) The brome mosaic virus-based recombination vector triggers a limited gene silencing response depending on the orientation of the inserted sequence. *Arch. Virol.* **155**, 169–179.
36. Pflieger, S., Blanchet, S., Camborde, L., Drugeon, G., Rousseau, A., Noizet, M., Planchais, S. and Jupin, I. (2008) Efficient virus-induced gene silencing in Arabidopsis using a 'one-step' TYMV-derived vector. *Plant J.* **56**, 678–690.
37. Qu, F., Ye, X., Hou, G., Sato, S., Clemente, T.E. and Morris, T.J. (2005) RDR6 has a broad-spectrum but temperature-dependent antiviral defense role in *Nicotiana benthamiana. J. Virol.* **79**, 15209–15217.
38. Ratcliff, F., Martin-Hernandez, A.M. and Baulcombe, D.C. (2001) Technical Advance. Tobacco rattle virus as a vector for analysis of gene function by silencing. *Plant J.* **25**, 237–245.
39. Roossinck, M.J., Sleat, D. and Palukaitis, P. (1992) Satellite RNAs of plant viruses: structures and biological effects. *Microbiol. Rev.***56**, 265–279.
40. Ruiz, M.T., Voinnet, O. and Baulcombe, D.C. (1998) Initiation and maintenance of virus-induced gene silencing. *Plant Cell* **10**,937–946.
41. Sambrook, J. and Russell, D.W. (2001) *Molecular Cloning: A laboratory Manual*, 3rd edn. 2100 pp. Cold spring harbor, NY: Cold spring harbor laboratory press.
42. Schardl, C.L., Byrd, A.D., Benzion, G., Altschuler, M.A., Hildebrand, D.F. and Hunt, A.G. (1987) Design and construction of a versatile system for the expression of foreign genes in plants. *Gene* **61**, 1–11.
43. Schwach, F., Vaistij, F.E., Jones, L. and Baulcombe, D.C. (2005) An RNA-dependent RNA polymerase prevents meristem invasion by potato virus X and is required for the activity but not the production of a systemic silencing signal. *Plant Physiol.* **138**, 1842–1852.
44. Senthil-Kumar, M. and Mysore, K.S. (2011) New dimensions for VIGS in plant functional genomics. *Trends Plant Sci.* **16**, 656–665.
45. Simon, A.E., Roossinck, M.J. and Havelda, Z. (2004) Plant virus satellite and defective interfering RNAs: new paradigms for a new century. *Annu. Rev. Phytopathol.* **42**, 415–437.
46. Tao, X. and Zhou, X. (2004) A modified viral satellite DNA that suppresses gene expression in plants. *Plant J.* **38**, 850–860.
47. Turnage, M.A., Muangsan, N., Peele, C.G. and Robertson, D. (2002)

Geminivirus-based vectors for gene silencing in Arabidopsis.*Plant J.* **30**, 107–114.

48. Unver, T. and Budak, H. (2009) Virus-induced gene silencing, a post transcriptional gene silencing method. *Int. J. Plant Genomics***2009**, 198680.

49. Vaghchhipawala, Z., Rojas, C.M., Senthil-Kumar, M. and Mysore, K.S. (2010) Agroinoculation and agroinfiltration: simple tools for complex gene function analyses. *Methods Mol. Biol.* **678**, 65–76.

50. Vaistij, F.E. and Jones, L. (2009) Compromised virus-induced gene silencing in RDR6-deficient plants. *Plant Physiol.* **149**,1399–1407.

51. Vaistij, F.E., Jones, L. and Baulcombe, D.C. (2002) Spreading of RNA targeting and DNA methylation in RNA silencing requires transcription of the target gene and a putative RNA-dependent RNA polymerase. *Plant Cell* **14**, 857–867.

52. Valentine, T., Shaw, J., Blok, V.C., Phillips, M.S., Oparka, K.J. and Lacomme, C. (2004) Efficient virus-induced gene silencing in roots using a modified tobacco rattle virus vector. *Plant Physiol.* **136**, 3999–4009.

53. Verwoerd, T.C., Dekker, B.M. and Hoekema, A. (1989) A small-scale procedure for the rapid isolation of plant RNAs. *Nucleic Acids Res.* **17**, 2362.

54. Vijayapalani, P., Chen, J.C., Liou, M.R., Chen, H.C., Hsu, Y.H. and Lin, N.S. (2012) Phosphorylation of bamboo mosaic virus satellite RNA (satBaMV)-encoded protein P20 downregulates the formation of satBaMV-P20 ribonucleoprotein complex. *Nucleic Acids Res.* **40**, 638–649.

55. Vogel, J.P., Garvin, D.F., Mockler, T.C., Schmutz, J., Rokhsar, D., Bevan, M.W., Barry, K., Lucas, S., Harmon-Smith, M., Lail, K., Tice, H.,Grimwood, J., McKenzie, N., Huo, N.X., Gu, Y.Q., Lazo, G.R., Anderson, O.D., You, F.M., Luo, M.C., Dvorak, J., Wright, J., Febrer, M.,Idziak, D., Hasterok, R., Lindquist, E., Wang, M., Fox, S.E., Priest, H.D., Filichkin, S.A., Givan, S.A., Bryant, D.W., Chang, J.H., Wu, H.Y., Wu, W., Hsia, A.P., Schnable, P.S., Kalyanaraman, A., Barbazuk, B., Michael, T.P., Hazen, S.P., Bragg, J.N., Laudencia-Chingcuanco, D., Weng, Y.Q., Haberer, G., Spannagl, M., Mayer, K., Rattei, T., Mitros, T., Lee, S.J., Rose, J.K.C., Mueller, L.A., York, T.L., Wicker, T., Buchmann, J.P., Tanskanen, J., Schulman, A.H., Gundlach, H., de Oliveira, A.C., Maia, L.D., Belknap, W., Jiang, N.,Lai, J.S., Zhu, L.C., Ma, J.X., Sun, C., Pritham, E., Salse, J., Murat, F., Abrouk, M., Bruggmann, R., Messing, J., Fahlgren, N.,Sullivan, C.M., Carrington, J.C., Chapman, E.J., May, G.D., Zhai, J.X., Ganssmann, M., Gurazada, S.G.R., German,

M., Meyers, B.C.,Green, P.J., Tyler, L., Wu, J.J. Thomson, J., Chen, S., Scheller, H.V., Harholt, J., Ulvskov, P., Kimbrel, J.A., Bartley, L.E., Cao, P.J.,Jung, K.H., Sharma, M.K., Vega-Sanchez, M., Ronald, P., Dardick, C.D., De Bodt, S., Verelst, W., Inze, D., Heese, M., Schnittger, A.,Yang, X.H., Kalluri, U.C., Tuskan, G.A., Hua, Z.H., Vierstra, R.D., Cui, Y., Ouyang, S.H., Sun, Q.X., Liu, Z.Y., Yilmaz, A., Grotewold, E.,Sibout, R., Hematy, K., Mouille, G., Höfte, H., Pelloux, J., O'Connor, D., Schnable, J., Rowe, S., Harmon, F., Cass, C.L., Sedbrook, J.C., Byrne, M.E., Walsh, S., Higgins, J., Li, P.H., Brutnell, T., Unver, T., Budak, H., Belcram, H., Charles, M., Chalhoub, B. andBaxter, I. (2010) Genome sequencing and analysis of the model grass *Brachypodium distachyon*. *Nature* **463**, 763–768.

56. Voinnet, O. (2001) RNA silencing as a plant immune system against viruses. *Trends Genet.* **17**, 449–459.
57. Voinnet, O. and Baulcome, D.C. (1997) Systemic signalling in gene silencing. *Nature* **389**, 553.
58. Wang, X.B., Wu, Q., Ito, T., Cillo, F., Li, W.X., Chen, X., Yu, J.L. and Ding, S.W. (2010) RNAi-mediated viral immunity requires amplification of virus-derived siRNAs in *Arabidopsis thaliana*. *Proc. Natl Acad. Sci. USA* **107**, 484–489.
59. Yang, C.C., Liu, J.S., Lin, C.P. and Lin, N.S. (1997) Nucleotide sequence and phylogenetic analysis of a bamboo mosaic potexvirus isolate from common bamboo (*Bambusa vulgaris* McClure). *Bot. Bull. Acad. Sin.* **38**, 77–84.
60. Yang, C.D., Liao, J.T., Lai, C.Y., Jong, M.H., Liang, C.M., Lin, Y.L., Lin, N.S., Hsu, Y.H. and Liang, S.M. (2007) Induction of protective immunity in swine by recombinant bamboo mosaic virus expressing foot-and-mouth disease virus epitopes. *BMC Biotechnol.* **7**,62.
61. Yuan, C., Li, C., Yan, L., Jackson, A.O., Liu, Z., Han, C., Yu, J. and Li, D. (2011) A high throughput barley stripe mosaic virus vector for virus induced gene silencing in monocots and dicots. *PLoS ONE* **6**, e26468.
62. Zhang, C., Yang, C., Whitham, S.A. and Hill, J.H. (2009) Development and use of an efficient DNA-Based viral gene silencing vector for soybean. *Mol. Plant-Microbe Interact.* **22**, 123–131.
63. Zuker, M. (2003) Mfold web server for nucleic acid folding and hybridization prediction. *Nucleic Acids Res.* **31**, 3406–3415.

Citations

CHAPTER 1

Cruz VMV, Kilian A, Dierig DA (2013) Development of DArT Marker Platforms and Genetic Diversity Assessment of the U.S. Collection of the New Oilseed Crop Lesquerella and Related Species. PLoS ONE 8(5): e64062. doi:10.1371/journal.pone.0064062

CHAPTER 2

Lane MA, Kimber M, Khokha MK (2013) Breeding Based Remobilization of Tol2 Transposon in Xenopus tropicalis. PLoS ONE 8(10): e76807. doi:10.1371/journal.pone.0076807

CHAPTER 3

Yang X, Wang F, Su J, Lu B-R (2012) Limited Fitness Advantages of Crop-Weed Hybrid Progeny Containing Insect-Resistant Transgenes (Bt/CpTI) in Transgenic Rice Field. PLoS ONE 7(7): e41220. doi:10.1371/journal.pone.0041220

CHAPTER 4

Dalecky A, Ponsard S, Bailey RI, Pélissier C, Bourguet D (2006) Resistance Evolution to Bt Crops: Predispersal Mating of European Corn Borers . PLoS Biol 4(6): e181. doi:10.1371/journal.pbio.0040181

CHAPTER 5

Yi D-K, Kim K-J (2012) Complete Chloroplast Genome Sequences of Important Oilseed Crop Sesamum indicum L. PLoS ONE 7(5): e35872. doi:10.1371/journal.pone.0035872

CHAPTER 6

Qaim M, Kouser S (2013) Genetically Modified Crops and Food Security. PLoS ONE 8(6): e64879. doi:10.1371/journal.pone.0064879

CHAPTER 7

Visser JC, Bellstedt DU, Pirie MD (2012) The Recent Recombinant Evolution of a Major Crop Pathogen, Potato virus Y. PLoS ONE 7(11): e50631. doi:10.1371/journal.pone.0050631

CHAPTER 8

Jingjing Yang, Yuanyuan Wang, Huolin Shen and Wencai Yang. 'In Silico Identification and Experimental Validation of Insertion–Deletion Polymorphisms in Tomato Genome" doi: 10.1093/dnares/dsu008

CHAPTER 9

D. A. Vaughan, E. Balázs and J. S. Heslop-Harrison. From Crop Domestication to Super-domestication. Ann Bot (2007) 100 (5): 893-901.doi: 10.1093/aob/mcm224

CHAPTER 10

Ming-Ru Liou, Ying-Wen Huang1, Chung-Chi Hu, Na-Sheng Lin andYau-Heiu Hsu A dual gene-silencing vector system for monocot and dicot plants DOI: 10.1111/pbi.12140

INDEX